Coltan

D0870649

For my mother, Denise Nest (1938–2009),
who loved the excitement of a new research project

Coltan

MICHAEL NEST

polity

Copyright © Michael Nest 2011

The right of Michael Nest to be identified as Author of this Work has been asserted in accordance with the UK Copyright, Designs and Patents Act 1988.

First published in 2011 by Polity Press

Polity Press
65 Bridge Street
Cambridge CB2 1UR, UK

Polity Press
350 Main Street
Malden, MA 02148, USA

All rights reserved. Except for the quotation of short passages for the purpose of criticism and review, no part of this publication may be reproduced, stored in a retrieval system, or transmitted, in any form or by any means, electronic, mechanical, photocopying, recording or otherwise, without the prior permission of the publisher.

ISBN-13: 978-0-7456-4931-3
ISBN-13: 978-0-7456-4932-0(pb)

A catalogue record for this book is available from the British Library.

Typeset in 10.25 on 13 pt FF Scala
by Servis Filmsetting Ltd, Stockport, Cheshire
Printed by MPG Printgroup, UK

The publisher has used its best endeavours to ensure that the URLs for external websites referred to in this book are correct and active at the time of going to press. However, the publisher has no responsibility for the websites and can make no guarantee that a site will remain live or that the content is or will remain appropriate.

Every effort has been made to trace all copyright holders, but if any have been inadvertently overlooked the publisher will be pleased to include any necessary credits in any subsequent reprint or edition.

For further information on Polity, visit our website: www.politybooks.com

Contents

Abbreviations

CNDP	Congrès National pour la Défense du Peuple
DRC	Democratic Republic of the Congo
EITI	Extractive Industries Transparency Initiative
FDLR	Forces Démocratiques pour la Libération du Rwanda
MLC	Mouvement pour la Libération du Congo
MONUC	Mission de l'Organisation des Nations Unies au Congo
NGO	non-governmental organization
OECD	Organisation for Economic Co-operation and Development
PARECO	Patriotes Résistants du Congo
RCD-Goma	Rassemblement Congolais pour la Démocratie–Goma
RCD-ML	Rassemblement Congolais pour la Démocratie–Mouvement de Libération
SOMIGL	Société Minière des Grands Lacs
SOMINKI	Société Minière et Industrielle du Kivu

Figures, tables and text boxes

Figures

Tables

Text boxes

Acknowledgements

US Secretary of State, Hillary Clinton, once said 'it takes a village to raise a child', and I can attest that it also takes a village to write a book. Thank you to Polity Press for giving me the opportunity to write about this most modern metal, especially Louise Knight and David Winters who entrusted me with the job and gave me feedback on the manuscript.

I cannot thank Claudia Halbac enough for lending her intellect to the project, and for the repartee. Her advice on the manuscript made it so much better and opened a wonderful new chapter in our friendship. Julian Kelly also provided extensive editorial comments that immeasurably improved the manuscript – thank you for being so generous with your time. Evalynn Mazurski cast an eagle eye over the tables and figures, and over the course of the project listened with collegial patience and interest to each new development. Isaac Djumapili gave me firsthand insights into the life of a coltan trader and commented on the manuscript. Who would have thought I would meet an *Mbuti* coltan trader in Sydney? His stories, his wife's cooking and the courtesy of his children transported me right back to the Congo. Two anonymous reviewers also provided helpful advice.

Ian Redmond of the Gorilla Organization and Aloys Tegera of the Pole Institute in Goma kindly gave me permission to make use of the treasure trove of interviews that were conducted as part of research projects which they had organized in the early 2000s. These qualitative field data remain an important source of information on the lives, aspirations and fears of coltan miners. Aloys Tegera provided further clarification on several additional issues.

A special thank you to Emma Wickens and Ulric Schwela of the Tantalum-Niobium International Study Center in Brussels, who patiently answered my many emails, provided data and clarified aspects of the tantalum industry that I found difficult to understand.

Thank you to Broederlijken Delen in Belgium, Melbourne Zoo in Australia and Maurice Carney of Friends of the Congo in Washington, DC for permission to reproduce posters created for their coltan-related initiatives.

The participants in the focus group that I organized in Sydney helped me to understand the logic and motivations of activists interested in justice issues related to the exploitation of natural resources. I would very much like to name them, but promised anonymity – thank you for generously sharing your time, energy and ideas on a rainy day. Thank you as well to Deborah Boswell for her scribing, enthusiasm and support. Lucy Hobgood-Brown also provided advice on organizing the focus group and, most of all, has been a source of encouragement and enthusiasm.

I pursued many people and organizations for clarification of details. Thank you to Dan Bucknell, Dirk Küster, Elisabeth Wood, Estelle Levin, Frank Melcher, Gary Coussins, Harrison Mitchell, Jacques Batumike, Jill Dobson, Kevin Hobgood-Brown, Laura Seay, Lee Ann Fujii, Matthew Cuolahan, Mei Yueh Hsieh, Odile Ruijs, Patrice Nyembo, Philippe Le Billon, Renee Cramer, Richard Burt, Serena Fletcher and Séverine Autesserre, as well as Global Witness and Metal-Pages. Any errors are, of course, mine alone.

Sometimes there is serendipity in life, and so it has been with Ramin: a materials scientist, under the same roof, able to answer questions about ductility, alloys and industrial applications of metals, and a whiz with Adobe Photoshop. Thank you, once again, for the love, patience and support.

1

Facts, figures and myths

A decade ago virtually no one except geologists had heard of tantalite, or 'coltan' as it is known in the Congo. Today, it is discussed at the United Nations, in the media, at student teach-ins and on activist websites, and is linked to some of the worst atrocities to blight the planet – mass rape, slave labour, extrajudicial killings and the illegal arms trade. There is even *Coltan* the novel, and 'Coltan Rush', a 'groove against war' by the Afro-jazz band, Bantunani. Whereas coltan was once an obscure mineral, there is now contestation over how it is produced, traded and sold. A *politics* of coltan has been brought into existence, one that encompasses warlords, transnational corporations, determined activists, Hollywood film stars, the rise of China, and the latest iGadget from Apple Inc. How did this happen? Why did an obscure mineral achieve such infamy? This volume analyses the two issues that have come to define coltan politics: the relationship between coltan and ongoing violence in the Democratic Republic of the Congo (hereafter DRC or 'Congo'), and contestation to reshape the global coltan supply chain.

Coltan gained widespread attention from journalists, activists and social scientists in 2001, when reports by the United Nations linked it, and other minerals, to ongoing violence in the DRC. Armed groups waging war were reportedly seizing control of coltan mines to take advantage of the record high prices which the mineral was fetching on the international market. Militia leaders boasted to journalists about the millions of dollars they were making in profits. In a complicated and seemingly unending conflict in which causes and motivations seemed to change from year to year, this provided a moment of clarity: armed groups were fighting to control coltan deposits and using violence against civilians who got in the way, and coltan was being bought by multinational corporations and used to make electronic gadgets, such as mobile phones, laptops, iPods and personal digital assistants, for Western consumers . . . or so the story went.

Over the 2000s the coltan industry became a 'lightning rod' for those concerned about conflict in the Congo. Activists, frustrated by a general lack of interest, suddenly had a symbol they could use to link the public to violence in far-off Congo: the mobile phone or 'cell' phone. Ordinary citizens became implicated. A US Senator claimed 'without knowing it, tens of millions of people in the United States may be putting money in the pockets of some of the worst human rights violators in the world simply by using a cell phone or laptop computer.'[1] Governments held inquiries into allegations that corporations from their countries bought coltan that had passed through the hands of armed groups; the United Nations and non-governmental organizations (NGOs) kept publishing reports implicating more and more companies in the illegal trade of natural resources from the DRC; and activists urged consumers to boycott products made from 'conflict minerals'. There are some inaccuracies and exaggerations in NGOs' and activists' accounts of cause and effect about coltan and conflict, but there is a determination to politicize the exploitation of natural resources such as coltan because of its perceived relationship with inequity and violence.

Coltan or tantalum?

Coltan is an abbreviation of columbite-tantalite, a mixture of two mineral ores, and is the common name for these ores in eastern Congo. *Tantalum* is the name of the metal extracted from tantalite-bearing ores, including coltan, after processing. Scholars, journalists and activists who have sought to understand and publicize the relationship between the tantalum supply chain and violence in the Congo, popularized the term coltan because this is the name used in the DRC. 'Coltan' has subsequently become widely used outside industry and scientific circles. Where the term 'coltan' is used in this volume, it refers only to unprocessed tantalite-bearing ores from the DRC.[2] 'Tantalite' is used when referring to tantalum-bearing ores in general or from countries other than the Congo. 'Tantalum' is used when referring to the processed metal. This volume analyses the global tantalum industry to provide some context to the economic forces and international actors that shape demand for coltan, but the volume is primarily concerned with coltan from the Congo, not tantalite or tantalum more generally. The title of this volume reflects broad recognition of the term 'coltan' and the fact that the political story worth telling is about coltan.

Volume structure and arguments

The present chapter describes where tantalum is found and produced and what it is used for, and analyses why its price has fluctuated, looking particularly at the causes of the price spike in 2000. The chapter puts to rest some popular myths about coltan and tantalum. For example, the Congo is *not* the repository for 80 per cent of the world's tantalite reserves nor does it produce 80 per cent of the world's tantalite, as is often claimed. Tantalite is also not a special mineral subject to ever-increasing demand. It has economic characteristics similar to other commodities: price changes are cyclical, and supply and demand ebb and flow.

Chapter 2 analyses the organization of production, trade and markets. Most tantalite is extracted in modern industrial mines in politically stable conditions where property rights are secure, and is sold under long-term contracts to processors. By contrast, coltan is produced using artisanal (pick and shovel) and small-scale mining methods. These methods are used because of a combination of factors unique to the DRC: the timing of the price spike in 2000, the geological characteristics of coltan, cheap labour and, most importantly, the weakness of state institutions. The latter has resulted in unclear and insecure property rights, impunity for breaking contracts, and poor infrastructure. Whereas in other countries customary institutions such as chiefs enforce contracts and protect property in the absence of a strong state, in the DRC these institutions have been weakened by war. The ways in which production and trade of coltan are organized in the DRC has created specific opportunities for profit-making by armed groups, which have been attracted to coltan for this reason. Economic networks centred on Rwanda have been pivotal to both the legal and illegal trade in coltan out of the DRC to the international market.

Chapter 3 analyses the relationship between coltan and conflict in the DRC. It commences by arguing that, while natural resources have played a role in conflict, this is typical of other conflicts as well and does not make violence in the Congo distinctive. The chapter summarizes the five waves of violence that have afflicted eastern DRC since the early 1990s, and analyses the degree to which coltan and other minerals have been factors in belligerents' strategies and motivations. The DRC conflict is not a single-issue conflict, it is not primarily over natural resources, and the motivations of warring parties have evolved over time. While control of the minerals sector including coltan has been contested by armed groups, they have also fought over agricultural land, opportunities to tax trade more generally, political dominance, national security, ethnic grievances, and to settle scores. Relative to other minerals and other motivations, coltan is not that important as a factor causing violence, although in 2000 and 2001 it was more significant, due to high prices, than in subsequent years. There is also no direct causal link between coltan and violence against civilians, including the mass sexual violence that characterizes conflict in the DRC. The fact that armed groups that control coltan mines and tax the coltan trade also commit rape and murder is incontrovertible. However, there is no evidence that this violence would stop if coltan disappeared from the equation.

Chapter 4 traces the evolution of advocacy, campaigns and initiatives focused on reshaping the coltan and tantalum supply chains in order to end war in the Congo. It analyses the objectives of ten initiatives and the tactics they employ, which range from celebrity testimonies on YouTube, to boycotts by college students of mobile phones containing 'conflict minerals', to 'freeze actions' on city streets in the Netherlands, to pictures of gorillas on phone recycling campaign posters in Australia, to training in 'ethical mining' for Congolese miners. Seven global initiatives that focus on improving governance in extractive industries more generally, and therefore have some potential to reshape the coltan industry, are also discussed. The tactics adopted in coltan campaigns expose underlying differences of opinion about the most effective solutions to the conflict. The key difference is whether the priority should be cutting off coltan profits to armed groups, or building the capacity of the DRC state to better manage development of natural resources. The chapter analyses the challenges these campaigns face in a global activist 'marketplace' crowded with competing global justice issues. The chapter argues that the political significance of coltan lies not in its actual causal link to violence, but in activists' effective manipulation of mobile phones (which contain tiny quantities of tantalum) as a symbol of how ordinary people and multinational corporations far from Africa are implicated in the Congo's coltan industry and therefore its conflict.

Chapter 5 analyses the future of coltan politics. It argues that activist campaigns have led to changes in the tantalum supply chain, in that Western corporations are pulling out of the market for coltan, causing a reconfiguration of the industry. However, this has stopped neither the trade in coltan nor profits filtering back to armed groups. China has become the chief destination for coltan as Chinese firms carve out a role in the global minerals, metals and manufacturing industries. The future outlook for tantalum is positive, but demand for the end products containing it will increasingly come from markets in developing countries. There are three key lessons from the politics of coltan for the politics of natural resources more generally. First, companies should be aware that, no matter how innocuous they think their business is, with time and money activists can create and propagate a compelling narrative of injustice that implicates their firm. Second, while non-state actors such as NGOs and multilateral organizations have successfully carved out a role in natural resource politics, this role can never be assured, given the rise of new corporations and countries, especially China, in the global minerals and metals markets. Third, if Western activists want to maintain the influence they have fought so long and hard to achieve, they need to devise messages that resonate with consumers in developing countries, seek ways to pressure corporations whose domestic reputation is little affected by their actions overseas, and find methods of ensuring that governments, with a tradition of not engaging with domestic political issues in other countries, will care about what their corporate citizens are doing.

There are two issues that the volume does not explore. First, it does not discuss how revenue from coltan should be invested and distributed. Certainly resource revenues are of critical importance to economic growth and the alleviation of poverty in the Congo. However, a discussion of the appropriate management and use of coltan revenues requires a wider survey of the many natural resources found in the Congo (including cobalt, copper, diamonds, gold, manganese, oil, timber, tin, tungsten, uranium and water), not just coltan, and is therefore beyond the scope of this volume. Second, while the volume analyses coltan initiatives and in the conclusion weighs up whether a concerned citizen should engage in these, it is not a 'how to' manual for activists or governments seeking to restructure the coltan supply chain.

Due to its broad political scope, this project was informed by many different fields of scholarship. However, in an effort to keep the volume readable, many facts and scholarly arguments are not directly referenced in the text. Instead, readers wishing to explore in more depth the various discussions and arguments to which this volume refers should consult the *Selected readings* section.

All monetary amounts are in United States dollars. All measurements are in metric (kilograms and metric tonnes), except tantalum prices which are in dollars per pound (the market convention).

When the volume uses the term 'armed group', this includes the DRC army.

Properties and characteristics of tantalum

Tantalum is classified as a 'minor metal'. It is grey-blue through black in colour, has the symbol Ta in the periodic table of the elements and its atomic number is 73. Tantalum's metallurgical properties make it indispensable to hi-tech industries. It is an excellent conductor of electricity; it has a very high melting point of 3,017°C (5,463°F), even when conducting electricity; it is very strong; it can be easily shaped and formed into different products; and it is highly resistant to corrosion.

These characteristics make tantalum ideal for a range of metal products. Its ability to conduct electricity is useful for mobile communications devices; its high melting temperature enables alloys (mixtures of metals) containing tantalum to remain in a solid state in extremely high temperatures, rather than melt; and its anti-corrosive characteristics mean that it will not rust or be affected by acidic environments. Table 1.1 shows the proportion of global tantalum production used for various industrial applications.

About two-thirds of tantalum production is used in the electronic capacitor industry to make tantalum capacitors. In the same way that a dam stores water from a river and regulates its flow to crops for agriculture, capacitors store and regulate the flow of electricity from batteries, or other power source, to the parts of an electronic device that perform functions, such as the display windows of mobile phones or storage areas for digital information. Capacitors are widely used in digital electronic devices such as personal digital assistants, iPods, digital cameras, liquid crystal display screens, DVD players, gaming platforms such as Nintendo, Xbox and PlayStation2, laptops and mobile phones. Tantalum capacitors are ideal for these products because, even in the crammed space inside a mobile phone, for example, they will not cause surrounding metals and plastics to heat up and melt because of their good heat conduction properties. Furthermore, only tiny quantities of tantalum are required – the average mobile phone contains less than 20 milligrams of tantalum. Tantalum is partly

Table 1.1   Industrial applications for tantalum, 2005

Industrial application	Percentage of all uses
Capacitors, e.g. mobile phones, laptops, gaming platforms, iPods	68
Other electronic and optic, e.g. memory chips in personal computers and igniter chips for car airbags	11
Superalloys, e.g. jet engines, turbines, space vehicles, nuclear reactors, power plants and chemical equipment	8
Carbides (carbon-based mixture of elements), e.g. cutting tools, drill bits	5
Coatings, e.g. silicon wafers, optical devices, camera lenses	2
Ammunition, e.g. military and recreational ammunition	1
Other, e.g. ink jet printers, x-ray film, surgical instruments, hip replacements	3

Sources: Wilson, Lorimer (2007), *Tantalum – A tantalizing commodity investment opportunity.* 22 August 2007. Retrieved 6 April 2010 from http://www.safehaven.com/article/8246/tantalum-a-tantalizing-commodity-investment-opportunity; and Jeangrand, Joel (2005), *Comprehensive Strategic Analysis of the Tantalum Industry.* Master's thesis, Simon Fraser University, Vancouver, p. 41.

responsible for the miniaturization of electronic devices and for the growth of markets for these products.

There are capacitors made of other materials, including ceramics and aluminium. These types of capacitors are less expensive than tantalum capacitors. However, they can store less electricity per unit volume than tantalum. For example, to get equal performance from a telephone containing a tantalum capacitor on the one hand, and a telephone containing a ceramic capacitor on the other, the latter would have to be considerably larger – making it a much less 'mobile' phone.

Tantalum is a critical material in other industries, such as the aerospace, automotive, energy generation and medical devices industries, mainly in the form of alloys. Tantalum alloys are strong and have high temperature resistance to cracking. Their heat resistant properties are useful for jet and automotive engines, which get extremely hot when working but will cool down rapidly when the engine is off. Tantalum alloys are also found in orthopaedic implants and heart pacemakers. Tantalum's resistance to corrosion makes it useful for medical devices such as these, which need to be implanted in human tissue.

Prices, supply and demand

Tantalum ores traded on the international market contain the tantalum-oxide compound known as tantalum pentoxide, which has the chemical formula Ta_2O_5. Concentrates – ore that has been through initial sorting and processing – typically contain between 10 and 40 per cent tantalum pentoxide. The price of tantalum that is reported by metals marketing services is for tantalum pentoxide, and the price varies depending on the ore's tantalum pentoxide concentration.

Like other commodities, the price of tantalum is linked to supply and demand, and supply and demand is cyclical. Industrial innovation caused demand for tantalum to steadily increase over the twentieth century and into the 2000s, but there have still been peaks and troughs of demand and price, including two notable price spikes: in 1980 and, most recently, in 2000.

The first significant rise in demand occurred during the Korean War (1950–53), when it was discovered that tantalum was useful for military applications. The United States was a small producer of tantalum at that time, but foreseeing the metal's usefulness the US government launched a worldwide initiative to buy tantalum and develop a stockpile, causing a doubling of prices. The government stopped buying domestic ore for its stockpile in 1958, and this caused tantalum prices to halve – the first price boom and bust. In the 1960s

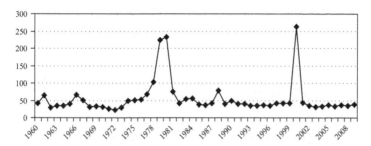

Source: Prices 1960–98: Cunningham, Larry D. (1998), Tantalum. In: *Metals Prices 1998*, United States Geological Survey (USGS), Washington DC, pp. 143–5. Prices 1999–2008: USGS, Washington DC (various years), Tantalum. In: *Minerals Yearbook.*

Figure 1.1 Tantalum price movements in 2007 dollars per pound, 1960–2008.[3]

tantalum became widely used in the industries for which it is known today: consumer electronics, chemicals, aerospace and machinery manufacturing. Prices fluctuated, however, as supply and demand never quite matched up. Prices declined from the late 1960s through to the early 1970s, but concerns about declining supply in the 1970s resulted in rapid price increases at the end of the decade. The years 1978–80 saw a price spike and crash. High prices during the boom caused industry to look for substitutes for tantalum, increased recycling of tantalum, and were the catalyst for exploration to identify new deposits. After declining in the mid-1980s, prices increased through to 1988, only to decline again into the early 1990s. Prices remained constant until 2000 when the price soared and 'coltan fever' took hold across eastern DRC, only to crash in 2001. Prices then remained steady for the rest of the decade.

Figure 1.1 shows movements in tantalum prices from 1960 to 2008. Prices reflect the year-end price of tantalum on the international spot market, and not the price of tantalum sold through forward contracts (see chapter 2 for a discussion of spot markets and forward contracts). Prices are in dollars per pound, expressed in 2007 dollars.

Price booms

Coltan came to the attention of the general media in 2001 when reports began filtering in of warlords in the Congo earning enormous profits from a rare mineral and a frenzied 'coltan rush' of miners into the jungle to exploit deposits. Links were made between the high prices tantalum was fetching on the international market, Western consumers' demand for the latest electronic gadget (the 'hot' item for Christmas 2000 was Sony's PlayStation2) and the profiting by armed groups from sales of coltan.

As shown in Figure 1.1, prices did indeed soar. In early 2000 tantalite ore was selling at between $30 and $40 per pound, by September 2000 it was selling for over $100 per pound, and by December 2000 it was selling for up to $300 per pound, depending on the concentration of tantalum pentoxide. By October 2001 prices were back to between $30 to $40 per pound.

The price boom was widely reported as being caused by a shortage of tantalum, which in turn was blamed for the scarcity of PlayStation2 before Christmas 2000. In fact, the price rise was caused more by speculation than by any real shortage of tantalum. What appears to have happened is this: in the late 1990s there was a surge in demand for consumer electronic devices. New products including second-generation mobile phones, gaming platforms and ultra-light laptops came onto the market and proved very popular. These devices used tantalum capacitors. The major tantalum processors, Cabot and H.C. Starck, knew this meant there would be future growth in markets for tantalum base products, especially tantalum powder which is used in the manufacture of capacitors. Nervous that they might lack sufficient stocks to feed this growth, Cabot and H.C. Starck placed large orders with producers via expensive long-term contracts. Much of the world's tantalum production became locked into contracts stretching into the 2000s. Processors that could not secure future supplies of tantalite risked losing their market share. Meanwhile the share price of companies that made capacitors and electronics products continued to rise, but there was suddenly very little tantalite available because supplies were already committed to certain processors. Demand rose strongly as supply on the open market became constrained, resulting in soaring prices driven by speculators and other traders trying to get their hands on tantalite available outside long-term contracts. The largest source of tantalite on the open or 'spot' market was coltan. Thanks to economic networks that had been established in 1998 and 1999 during the first years of the Congo War, minerals traders and military officials were perfectly placed to funnel it out of the country. The price of tantalum fell when buyers refused to pay the high prices, probably realizing that they were paying far too much for a mineral that was not in short supply and that they could not sell to processors, who had already committed themselves to buying more than enough tantalite. The flood of coltan from the Congo onto the international market contributed to the price drop.

The previous price spike, in 1978–80, had similar causes to that of 2000, but unlike the latter it did not cause a 'coltan rush'.[4] In the late 1970s there was a perceived looming shortage of tantalum supply – perceptions fed by civil war in Mozambique that caused the closure of its tantalite mines and a statement by the Tantalum Mining Corporation of Canada that reserves at its Lake Bernic mine in Manitoba were running down and more exploration worldwide was needed. The latter was misinterpreted by some buyers as meaning that supplies were about to terminate. Because of the perceived pending shortage, prices jumped in 1978 and then soared from $103 to $224 in 1979. The most readily available source of tantalite for the spot market at that time was from tin slag in Thailand, and there are stories of roads and buildings being demolished to get at the tin slag lying underneath. As in 2000, there was no real shortage and the market frenzy was caused largely by speculators. The perceived shortage was, however, the catalyst for new exploratory work by East German geologists on the tantalite deposits in Ethiopia and Mozambique that are mined today.

The reason the 1978–80 price spike did not cause a coltan rush in the Congo is due to a mix of institutional, geological and social factors. First, as a consequence of a stronger DRC state, property rights were more secure in the late 1970s. In the late 1970s tantalum pentoxide was extracted as a by-product of industrial tin mining, mostly by the DRC state-owned company, Société Minière et Industrielle du Kivu (SOMINKI), and not by artisanal or small-scale miners. Tantalite production from the DRC was significantly higher in 1980 compared to either 1978 or 1979, so SOMINKI was obviously able to take advantage of higher prices by increasing output. Higher output also suggests that SOMINKI was able to keep artisanal and small-scale miners off its concessions, possibly with support from the police (a more capable institution in the late 1970s compared to 2000) and the military. Second, because tantalite was produced as a by-product of industrially mined tin and not from alluvial deposits, artisanal and small-scale miners would have lacked the processing equipment required to extract tantalum pentoxide. They would have been unable to exploit tantalum found in cassiterite (tin ore) even if they obtained it. By contrast, in the late 1990s and 2000s, coltan produced in the DRC was extracted from deposits with higher concentrations of tantalum pentoxide requiring only rudimentary processing before sale. Finally, most Congolese had little knowledge of tantalum because there was no broad participation by the general population in tantalite mining prior to 2000. A Congolese former coltan trader commented 'in about 2000 we started to hear about coltan, but it was not known before that'.[5] As tantalite was extracted as an industrial by-product in processing plants, ordinary Congolese had neither the access nor the specialized education to know that what was ostensibly tin mining also produced quantities of an obscure mineral.

Inventories

The price boom in 2000 is linked to the creation of inventories. The size of inventories – that is, stockpiles of ore – has a major influence on prices. If producers know that processors have a small inventory they will try to charge more when selling on the spot market, because they will assume the processor will be keen to buy ore to keep their plant operating. Conversely, if buyers have a large inventory they will push for a lower price from producers.

One stockpile that was not accumulated by a commercial firm, but which has had a major influence on prices, was that of the US Defense Logistic Agency. As already mentioned, the US government decided to create a stockpile during the Korean War and in the early years of its conflict with the Soviet Union, when it feared that American industry would be unable to obtain sufficient quantities of critical metals. Following the break-up of the Soviet Union in 1991, the US government decided the stockpile was no longer required because it had become easy for American companies to obtain metals in an expanded free market that included former Soviet states. In 1993 it adopted the National Defense Authorization Act which set disposal targets for minerals previously stockpiled, and tantalum stocks were completely sold off in December 2007.

The largest processors of tantalite into tantalum base products, H.C. Starck and Cabot, built up inventories in the 2000s as they were committed to purchasing ore under long-term contracts. Capacitor manufacturers also built up inventories of tantalum powder that they had purchased from processors. By the end of 2008, processors and capacitor manufacturers had enough inventory to last until mid-2010, about 18 months. This was a good position to be in, given the drop in supply caused by the closure in December 2008 of the Wodgina mine in Australia, which at that time produced 30 per cent of the world's tantalite.

Sources, reserves and production

The global tantalum industry obtains tantalum from four sources: mines, recycled and scrap materials, inventories and the tailings from tin slag. The US Defense Logistics Agency's stockpile was a fifth source until it was sold off. Historically one of the most common sources of tantalum was the slag heaps around tin mines in Malaysia and Thailand, which contained between 2 per cent and 10 per cent tantalum pentoxide. In the 1980s about 40 per cent of the world's tantalum came from tin slag, but tantalum from these sources was largely depleted by the 1990s. Recycling will probably be a declining source of tantalum in the future due to the difficulty of separating out tantalum capacitors from increasingly miniaturized circuitboard chips in electronic devices. Figure 1.2 shows the percentage of tantalum from each source as of late 2007.

Tantalum sourced from mining is found in different ore mixtures around the world. The main mineral sources of tantalum in order of importance are tantalite, niobium, (columbite), tin slag, columbite-tantalite (coltan), and struverite – a tin-niobium-tantalum mixture. These tantalum-bearing ores are common throughout the world and are found

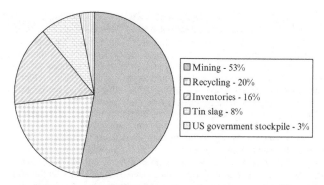

Mining - 53%
Recycling - 20%
Inventories - 16%
Tin slag - 8%
US government stockpile - 3%

Source: Zogbi, Dennis M. (2008), *Tantalum: Global market outlook: 2008–2012.* Paumanok Publications, Rockville, MD.

Figure 1.2 Sources of tantalum, 2007.

in more than forty countries. However, many deposits contain low concentrations of tantalum pentoxide making them not commercially viable for large-scale mining, although artisanal and small-scale miners are sometimes able to scrape a living from them.

Many media, scholarly and NGO reports state that tantalite is rare and that most tantalum comes from the DRC. The figure of 80 per cent is routinely used to describe the proportion of the world's tantalum that comes from the Congo, the proportion of the world's reserves found in the Congo, or the proportion of the world's reserves found in Africa. The 80 per cent figure is rarely referenced, and when it is the reference is unreliable. All data available indicates that these claims are false. See Box 1.1 for more information on the '80 per cent claim'.

The truth is that good data on tantalite reserves in the DRC is not available because good geological exploration work has not been done for at least two decades. After the early 1990s most mineral exploration companies were deterred from

Box 1.1   On the trail of '80 per cent'

According to the Tantalum-Niobium International Study Center in Brussels, the earliest reference to the Congo or Africa having 80 per cent of the world's tantalum reserves was probably a story from Agence France Presse around 1996. The earliest article the author could locate was from Britain's *Guardian* newspaper on 4 March 2001. The article states: 'Africa has 80 per cent of the world's supplies, and of these, three-quarters are found in the Kahuzi Biéga park and neighbouring land,' which places 60 per cent of global tantalum reserves in the DRC. The following month, on 7 April 2001, an equally reputable magazine, *New Scientist,* published an article repeating a version of this claim: that the DRC has 64 per cent of the world's tantalum reserves. Then, on 1 August 2001, an article on the *BBC News* website stated that the Congo has 80 per cent of the world's reserves of tantalum. By the end of 2001, newspapers, television news programmes, scholars and activists everywhere were saying that the Congo had 80 per cent of the world's tantalum. These claims are false, but the 80 per cent figure had become urban myth.[6]

Table 1.2   Myth busters: some facts about coltan and tantalum

Myth	Fact
1. There is a global shortage of tantalum	Tantalum is not uncommon and is found in many countries throughout the world
2. The Congo has 80% of global tantalum reserves	The precise size of the Congo's tantalum reserves is difficult to calculate. The most informed estimate is that Central Africa has around 9 per cent of global reserves. The DRC's reserves are the major component of these – perhaps 7 to 8 per cent of global reserves
3. The Congo is the world's largest producer of tantalum	Reliable production data are not available. However, for most of the 2000s the Congo may have produced around 20 per cent of the world's tantalite. Historically the largest producer has been Australia
4. There is an ever-increasing demand for tantalum	Overall demand has increased steadily since the Second World War, but it continues to be cyclical with peaks and troughs similar to other minerals

operating in the DRC due to violence and political instability. Table 1.2 contains some 'myth busters' about coltan and tantalum.

Data concerning tantalum reserves for different parts of the world is important because of what it tells us about the capacity of DRC coltan production to influence global markets. The DRC has two large deposits in terms of area: the Kibaran deposit running along a north-south axis through eastern DRC and into northern Katanga province, and the Eburnean deposit running along an east-west axis across Equateur and Orientale provinces in northern DRC. Most Congolese coltan comes from the Kibaran deposit in North Kivu and South Kivu. Rwanda also has deposits of tantalite, although these are far smaller than the DRC's. The fact that Rwanda produces tantalite makes it easy for Rwandan-based exporters to mix tantalite from the Congo and Rwanda and plausibly claiming that the ore all comes from Rwanda. While total Rwandan export volumes of tantalite clearly do not all originate from Rwanda, it is difficult to prove the origin(s) of any single consignment of ore. Tantalum ore 'fingerprinting' technology being developed in Germany may help to solve this problem (see chapter 4).

Calculating tantalite reserves in the DRC or any other country is not straightforward. The basic problem is that minerals lie in the ground, mostly out of sight. Reserves for any particular country also change from year to year depending on how much tantalum has been produced (depleting available reserves) and how many more deposits have been discovered. To calculate reserves, geologists use technology to determine the concentration of minerals in ore samples taken from a deposit, such as the concentration of tantalum pentoxide. Using information about the probable size of a deposit, an estimate of probable reserves is then developed. While calculating reserves involves estimations, these are based on science developed over a century, not on guesses.

The most recent and comprehensive data on reserves in Central Africa, including the DRC, Burundi, Rwanda and Uganda, is published by the Tantalum-Niobium International

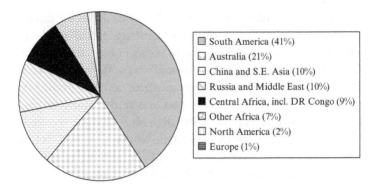

South America (41%)
Australia (21%)
China and S.E. Asia (10%)
Russia and Middle East (10%)
Central Africa, incl. DR Congo (9%)
Other Africa (7%)
North America (2%)
Europe (1%)

Source: Created by author using data from Burt, Richard (2010), Tantalum – A rare metal in abundance? *T.I.C Bulletin* 141, p. 5.

Figure 1.3 Global tantalite reserves, 2009.

Study Center in Brussels – see figure 1.3. Most of the 'Central Africa' reserves are from the DRC.

Production

Simply the fact that a country has tantalite deposits does not mean that these will be mined. Whether the ores are mined is a commercial decision that is made by the companies that own the mining rights, based on the cost of getting the ore to market and the price it will receive. The cost of mine development and mine operations includes fees and licences that are payable to governments and landowners; bringing electricity to the location, such as through power lines, or building a hydro-electric dam; bringing water to the site; design, engineering and construction of the mine; labour, including food and accommodation and, depending on the mine, social costs such as education and health clinics for staff and communities affected by the mining operations; transportation routes for ore, such as roads, railways, river barges or specialist ports to load ore onto ships; and compliance, such as environmental mitigation works, occupational health and safety measures and remedial work when the mining has been completed. The most important factor is the price that the mining company can obtain for its ore, and this depends on worldwide demand for tantalum base products. Critically, price and costs are affected by exchange rates, especially the value of the US dollar, as many firms sign contracts in dollars and tantalum ore is often bought and sold using dollars. Some mines cost hundreds of millions of dollars to develop, and it may take the mining company several years before the mine turns a profit. However, if operational costs (such as labour, electricity and transportation) remain steady once the mine becomes profitable, the returns can be very rewarding. Conversely, some variables can result in a mine becoming unprofitable, such as exchange rate fluctuations, price declines and escalating labour costs.

The number of large-scale mines producing tantalum using industrial methods is small, but there are also deposits that are exploited by artisanal and small-scale miners. Developed tantalite mines are found on all continents except Antarctica. Figure 1.4 shows a map of significant developed and prospective tantalite mines. Some of the developed mines in figure 1.4 were 'mothballed' on care and maintenance programmes and were no longer productive in late 2009 and early 2010.

Data available for global production of tantalite shows that primary production from mining rose steadily from the mid-1990s, almost doubled from 1998 to 2001 when it peaked at 1,692 tonnes, then fluctuated around the 1,600 tonne mark during the mid-2000s – see figure 1.5.

Data on primary production of tantalite by country is available although, like the data on reserves, figures are not entirely reliable for central African countries, for two main reasons. First, data is not systematically collected for production from

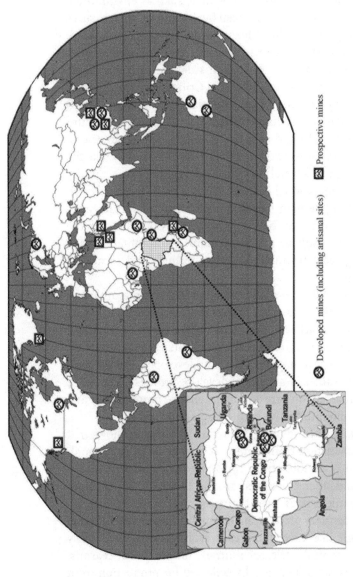

Source: Created by author. World map used with permission from University of Alabama; D.R. Congo map used with permission from International Crisis Group.

Figure 1.4 Locations of significant developed and prospective tantalite mines.

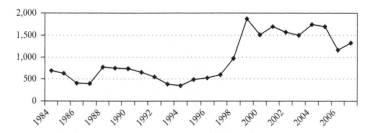

Source: Tantalum-Niobium International Study Center.

Figure 1.5 Estimated global tantalite primary production, 1984–
2007, metric tonnes.

artisanal and small-scale mines in the DRC where coltan is
produced. Second, production data that is available is based
on export statistics, and exports are under-reported because
individuals, companies and government officials involved
in the illicit coltan trade want to evade taxes. Rwanda's tantalite export data includes a large proportion of tantalite that
originates from the DRC, possibly half, by one estimate.[7]
Coltan that is smuggled from the DRC into Rwanda becomes
counted in Rwandan statistics when it is officially exported to
Europe or Asia.

Time series data for primary production of tantalite is
available for Africa from 1997 to 2007. Figure 1.6 reveals
Rwanda's theft of coltan from the DRC in 1999 and 2000,
when Rwandan 'production' increases due to its exportation
of coltan stolen from the Congo, and the DRC's coltan production boom in 2000, followed by the crash in early 2001 and
subsequent increase in 2002. It also shows new Mozambican
production coming on stream in 2003 and Ethiopian production that increased six-fold over the decade. The surge
in Zimbabwean production in 2002 is curious. The US
Geological Survey reported this production as being 'mine
output', but it is certainly an anomaly if this is true. Zimbabwe

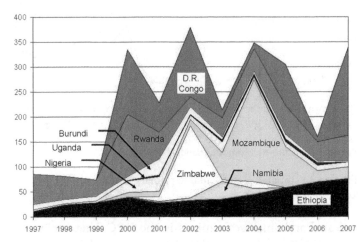

Sources: Figure reproduced with permission from Dr Frank Melcher of the German Federal Institute for Geosciences and Natural Resources. Originally published by Melcher, Frank, Graupner, Torsten, Henjes-Kunst, Friedhelm, Oberthür, Thomas, A. Sitnikova, Maria, Gäbler, Eike et al. (2008), *Analytical Fingerprint of Columbite-Tantalite (Coltan) Mineralisation in Pegmatites – Focus on Africa.* Paper for the Ninth International Congress for Applied Mineralogy, Brisbane, 8–10 September. The figure was updated by Dr Melcher using data from the United States Geological Survey and the Tantalum-Niobium International Study Center. Data for the Congo are from the DRC government's Centre d'Evaluation, d'Expertise et de Certification des Substances Minérales Précieuses et Semi-précieuses.

Figure 1.6 African tantalite production, 1997–2007, metric tonnes.

was one of the DRC government's military allies during this period and it is possible that tantalite was being transported from the Congo to Zimbabwe, where it was counted as Zimbabwean production. However, there were no coltan mines under DRC government control in 2002. Another possibility is that coltan from the DRC was being transported via Rwanda to Zimbabwe, but this is unlikely due to limited trade between the two countries in 2002 as a result of them being on opposing sides during the Congo War.

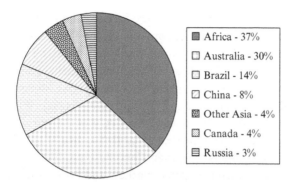

Source: Wallwork, Paul (marketing manager, Talison Incorporated) (2008), *Challenges facing the tantalum industry.* Presentation for the Minor Metals and Rare Earths conference, Hong Kong, 3–5 September 2008.

Figure 1.7 Sources of tantalite primary production, 2008.

According to the data in figure 1.6, in 2007 the DRC produced the majority of African tantalite (about 52%), followed by Ethiopia (22%), Rwanda (15%), Mozambique (7%) and Nigeria (3%). However, as mentioned, half Rwanda's official tantalite production may come from the DRC. Assuming this is true, it means about 60 per cent of African tantalite came from the DRC in 2007.

Figure 1.7 shows estimated global tantalite primary production by country or region for 2008. It shows that Australia was by far the largest single producer of tantalite in 2008, with all of its production coming from just one mine, Wodgina. The more interesting story, however, lies behind the 37 per cent of tantalite that is produced in Africa and what this might tell us about Congolese production.

Using the data in figures 1.6 and 1.7 and by making some assumptions, it is possible to estimate the Congo's share of global tantalite production. Assume that (1) the proportion of tantalite produced by each African country in 2007 as shown in figure 1.6 remained the same for 2008, when Africa produced 37 per cent of world production, and (2) that half of Rwandan tantalite actually comes from the DRC. Using these assumptions, the top country producers in 2008 in order would be: Australia (30%), DRC (21%), Brazil (14%), China (8%), Ethiopia (8%), Canada (4%), Rwanda (3%), Mozambique (3%), Russia (3%), Nigeria (1%) and 'other' (4%). While the DRC produced nowhere near 80 per cent of the world's tantalite, it may have been the second largest source of tantalum in the 2000s. If this is true, then Congolese production has the potential to destabilize global markets and cause price volatility due to the volume of coltan being sold on the international spot market outside long-term contracts.

Tantalite production continues to evolve along with the global economy, and production patterns have radically changed. From December 2008, mines in Australia, Canada and Mozambique were taken out of production and placed on care and maintenance programmes as the global financial crisis reduced global demand and prices. In the case of Australian tantalite, a strengthening Australian dollar and escalating labour costs (due to demand from China there has been an ongoing mining boom for Australian minerals resulting in ever-increasing salaries) probably made the Wodgina mine unprofitable. As a result of these closures, the DRC may have produced as much as one-third of the world's tantalite for a short period at the end of the 2000s, but this would have been a historical anomaly. Patterns of tantalite production are set to change once again in 2010, with the easing of the global financial crisis, increasing demand for tantalum and an announcement by Talison Inc. that it plans to reopen the Wodgina mine.[8] Conditions sufficiently favourable to allow the reopening of Wodgina would also result in the reopening of mines in Canada and Mozambique, causing the DRC's share of global production to shrink.

The closure and then reopening of Talison's Wodgina mine

brings together several threads of the tantalum story, linking demand, prices, inventories and calculations about production. Wodgina's closure in December 2008 was remarkable: 30 per cent of the world's tantalum production was shut down in one go. The equivalent in the oil world would be if Saudi Arabia and Russia both suddenly decided to turn off the oil production tap. Talison gave two reasons for the closure. First, as a result of the global financial crisis, demand for electronics products was lower, resulting in depressed demand for tantalite. Second, cheap coltan from Central Africa was undercutting the price of tantalite from Wodgina and Talison could no longer compete. This story of 'conflict coltan' making an ethical producer, Talison, go out of business has been latched onto by some analysts as evidence of the importance of Congolese tantalite to world markets and the future of the electronics industry.

The truth is less clear and more complex. Talison signed long-term forward contracts in the late 1990s before the price boom, and consequently missed out on earning the higher prices of 2000. Over the 2000s Talison's labour costs, like those for mining companies operating elsewhere in Australia, skyrocketed due to that country's mining boom. Tantalite from Wodgina that was profitable in the early 2000s was much less profitable five years later due to higher production costs. In mid-2008 Talison was reportedly negotiating for a large price increase from processors, perhaps 80 per cent, but processors refused: demand for tantalum base products was depressed because of the global financial crisis, they had accumulated large inventories during their long-term contracts and they saw no looming shortage of supply.[9] Talison then called the processors' bluff by closing down production.

The closure of Wodgina and the consequent withdrawal of a very large proportion of tantalum production from the global market has a precedent. Following the price spike of 1978–80, the Tantalum Mining Corporation's mine at Lake Bernic – then the largest producer of tantalum in the world – closed in similar circumstances. High prices had eventually resulted in supply outstripping demand and, once buyers realized there was no real shortage, the price of tantalum crashed (from $234 in 1980 to $76 in 1981). The Tantalum Mining Corporation was 'unable to sell a pound of concentrate', no doubt because processors wanted to use up the inventories they had accumulated. The company closed the Lake Bernic mine, which stayed shut until 1988, by which time Wodgina and another large Australian lithium-tantalite mine (Greenbushes) had commenced production. The Lake Bernic mine was reopened for the same reasons Talison decided to reopen Wodgina: prices had increased and the company wanted to reclaim its market share. This cycle was repeated in April 2009 when the Lake Bernic mine was again closed in response to reduced demand associated with the global financial crisis.

Mines are expensive to open and to operate, so placing a mine onto a care and maintenance programme is not done lightly. In particular, labour needs to be retrenched and because of this there is often strong political opposition to closure. Most industrial mines continue to operate through price lulls, even if production is unprofitable, because mining companies take a long-term view of production and profitability. However, long periods of reduced prices can result in a mine being temporarily closed, with the mine being reopened when future demand looks bright. Reopening a mine involves repairing degraded infrastructure and recruiting a new labour force – significant challenges.

After Wodgina's closure in 2008, processors turned to their inventories and waited out the global financial crisis. Through 2009 Asian demand for electronic and industrial products barely dimmed – and Chinese demand grew – and by early 2010 Japan, North America and Europe seemed to be pulling out of recession. Talison scented that demand for tantalite was likely to increase and scheduled Wodgina's reopening for early 2011. Talison was probably also cognisant that with the development of Egypt's Abu Dabbab mines, tentatively scheduled to open in 2013, it needed to get back into the market and sign more contracts if it wanted to maintain its market share. Processors were also getting jittery about a potential shortage in supply, given the US government's disposal of its stockpile and a shortage of recyclable scrap material (because of reduced capacitor manufacture), and may have had talks with Talison about resuming tantalite production. Even with the reopening of Wodgina and the opening of Abu Dabbab, demand is likely to exceed supply and tantalite prices are set to rise in the early 2010s.

Conclusion

Tantalite has economic characteristics similar to other minerals. Prices are cyclical, supply and demand ebb and flow, a shortage of supply drives exploration for new sources and efforts to find substitute materials, and mines are closed and reopened depending on fluctuations in international prices. Confirmed Congolese reserves of tantalite are dwarfed by those of Australia and Brazil, but this is because they are poorly quantified rather than small. The geological research that has been undertaken on the Kibaran deposit in eastern DRC suggests it may have the potential to be a major source of the world's tantalite, but this has not yet been proven.

The broad forces of the international minerals and metals market continue to influence the coltan industry in the Congo, but the price spike of 2000 started something different and distinctive. Tens of thousands of Congolese, in desperate economic straits because of conflict, moved en masse to coltan deposits and started to dig. A portion of these people left when the price dropped, but the effects of this coltan rush endure today: control by armed groups, new mining communities in national parks or where previously there were village societies organized around agriculture, and economic trading networks comprising Congolese civilians, private companies and military forces. Understanding how these actors in these communities organize production and trade of coltan in conditions of political instability, violence and a weak state is critical to understanding the politics of coltan.

2

Organization of production and markets

The popular image of tantalum is that it is hacked from the ground in chaotic mines lorded over by a thug with a gun. Journalists' reports, of travelling by motorbike and then on foot through the jungle to finally come across apocalyptic scenes of men and women working up to their waists in mud, have fed these images. The appearance of chaos is deceptive on two counts. First, most of the world's tantalum by quantity and value is produced in large industrial mines using modern methods requiring safety and environmental standards. Miners drive or take the bus to work, and return to quiet homes at the end of a shift. Second, where tantalum is produced by artisanal and small-scale methods, as is the case in the DRC, production is *organized*. The physicality of artisanal mining – mud, water, dirt, dust – should not be confused with an absence of order.

Artisanal and small-scale methods are used to mine coltan in eastern DRC because of the timing of the coltan rush in 2000 to exploit coltan deposits, the geological characteristics of Congolese coltan (compared to other tantalite-bearing ores), the low wage economy and, most importantly, the weakness of the Congolese state. The weak Congolese state is manifested in poor infrastructure, uncertain property rights caused by a capricious and corrupt justice system, and personal insecurity caused by the presence of anti-government militias and ill-disciplined and corrupt armed forces. These three factors have resulted in Congolese preferring micro-investments in equipment and stock, and therefore a preference for micro-production and trade. Markets in commodities are usually underpinned by the state's enforcement of contracts and security of property. In some countries where the state is weak, alternative community-based networks might provide protection of property and punish people who break contracts. In the DRC, community-based networks have been so destabilized by conflict – with customary chiefs being killed or fleeing, or having their authority challenged by young men with guns – that they struggle to perform this function.

The artisanal methods used in the Congo are a critical factor in armed groups being able to profit from coltan, because of the ease with which artisanal production can be controlled using violence and the opportunities for taxation created by multiple exchanges. One of the issues this chapter explores is why, given the importance of stability for optimal exploitation of coltan, there is violence around mine sites. There is an ethnic dimension to this violence, but ethnic networks also provide some security of exchange and are attractive to miners and traders for this reason.

To fully understand the politics of coltan it is necessary to understand how the production and supply chain is organized, and understanding how the supply chain is organized is essential to understanding how initiatives to regulate coltan are supposed to work. The tantalum supply chain involves several basic steps, depending on the final product. In the case of mobile phones it involves seven steps: (1) extraction of tantalite ore, (2) trade and export of ore to processing firms, (3) smelting and refining to create tantalum powder, (4) manufacturing of tantalum capacitors, (5) manufacturing of circuitboards, (6) manufacturing of the mobile phone, and (7) retail sale to consumers. It is at the end of the second step that coltan enters the global minerals market.

The organization of steps one and two is significant to

the politics of coltan because the way coltan is produced and traded creates opportunities for armed groups, and these steps are the focus of this chapter. Steps three to seven are significant for different reasons. The way these steps are organized is of some analytical interest due to the way processing, manufacturing and consumption have become specialized and globalized, features that complicate proposed certification schemes for minerals and challenge activists. However, it is contestation to influence the *participants* in these steps that is at the core of the international dimension of coltan politics. The transnational corporations that use components containing tantalum, and the consumers who buy the final end products, are the focus of both efforts by activists to draw attention to corporate involvement in the DRC war economy and by NGOs, governments and multilateral organizations to regulate coltan. Steps three to six are therefore discussed only briefly in this chapter, but are revisited in chapter 5. The final step in the tantalite supply chain, retail sale of consumer goods that contain coltan, is also discussed in chapter 5, which focuses on future markets for products containing tantalum.

Organization of production

Most tantalum is produced in capital-intensive, modern, industrial, open pit mines in Australia, Brazil, China and Canada. In Africa several different mining methods are used. Modern industrial techniques are used to mine tantalum in Mozambique, and will be used in Egypt's two planned tantalum mines; semi-industrial techniques are used in Ethiopia; and artisanal and small-scale methods are used in the DRC, Nigeria and Rwanda.

Large-scale industrial mining of tantalite involves open pit mines, sometimes very large, and the use of sophisticated machinery, including tracked excavators (diggers), large haulage trucks and mechanized conveyor belts to move ore and tailings. Basic concentration of ore – breaking it down into small pieces – is carried out on site. Transportation of tantalite concentrate is by train or truck to ports where it is shipped in bulk to processors.

Artisanal and small-scale mines, on the other hand, are low-tech operations. Pits up to 6 metres deep are dug using hand tools such as picks, crowbars and shovels. Explosive blasting or high-pressure hoses are used to loosen the rock substrate beneath the soil. Diesel generators are needed to run high-pressure hoses. Water is critical to the mining process, as it is used to separate and wash away earth from ore in sluices or other vessels. For this reason mining often occurs near rivers and creeks or, alternatively, waterways are diverted to ore deposits.

Unlike some other minerals, tantalum's physical and chemical properties make it particularly suited to artisanal and small-scale mining. Because tantalum has high density (i.e. it is heavy like gold, not light like aluminium oxide) as well as high levels of chemical inertness, it is relatively easy to separate tantalite out from dirt and other ores using methods relying on gravity. Expensive chemical treatment is not required until the processing stage, when concentrated ores must be chemically transformed.

Several factors determine whether large-scale modern mining methods, or artisanal and small-scale mining methods, will be used to extract a mineral. The value of a mineral is critical to the viability of different mining methods. Minerals that are valuable in small quantities, such as diamonds, gold and coltan, can be profitably mined by individuals using basic methods if the mineral is found near the surface, such as in alluvial deposits in riverbeds. If found in large enough deposits, these minerals can also be mined profitably using industrial methods. By contrast, minerals such as coal, iron ore, copper and uranium are worthless in small quantities, even if they are easy to obtain. Profitable mining of such minerals depends on industrial techniques that produce enormous quantities of ore at a low unit cost. The method used to produce a mineral also depends on the availability and cost of capital, labour and infrastructure and the strength of property rights. Secure property rights are essential before a modern mining company will invest in site preparation and equipment. Artisanal and small-scale miners also prefer secure mining rights if they can get them, especially as their investments, while small, can be substantial relative to savings and income.

Unfortunately for artisanal mining communities, artisanal production is more vulnerable to control by an armed group. Unlike modern large-scale mining, artisanal mining requires no expertise in financial or labour management or sophisticated production processes. These factors facilitate control by armed groups that are unskilled in mining, as well as enabling taxation of concentrate by armed groups when it is transported and frequently exchanged. The vulnerability of artisanal and small-scale miners to extortion, taxation and violence by armed groups does not mean that there will be violence wherever these methods are used. The key factor determining the degree of violence at an artisanal mine is probably the strength of institutions that can be used to settle disputes, especially over property rights. In some cases religious or customary leaders might also play an important role in dispute resolution. In the DRC, some local military and policing authorities have capacity to control violence in mines. However, as will be discussed in chapter 3, where these authorities have this power they have sometimes used it to personally profit from production and trade rather than upholding the rule of law to enable civilians to mine.

Tantalite mining in the Congo

Tantalite was first identified in the Congo in 1910, and was extracted as a by-product of cassiterite (tin) mining until the 1990s. In 2000, eastern DRC became enveloped in 'coltan fever', akin to the gold rushes of the nineteenth century, which caused an influx of people to coltan mining areas. Tantalite was no longer a by-product; it was the objective. In 2009, there were twenty-three coltan mining sites in eastern DRC: fourteen in North Kivu and nine in South Kivu. Probably the largest coltan-producing mine is Bisie mine in Walikale, North Kivu, which also produces bauxite and cassiterite. In 2008, Bisie had a population of about 13,000 people.

Before the Congo War of 1998–2003, the DRC government operated a state-owned mining company, Société Minière et Industrielle de Kivu (SOMINKI), in eastern DRC. SOMINKI used modern mining methods and had a monopoly on concessions. It was a large producer of gold, diamonds and cassiterite (from which tantalum was extracted as a byproduct). SOMINKI halted operations during the war in 1996–1997 against former President Mobutu, was restarted by the new government of Laurent Kabila, only to cease operations again during the occupation of eastern DRC by Rwandan and Ugandan troops between late 1998 and 2003. Since the demise of SOMINKI's operations, coltan has been produced using only artisanal and small-scale mining methods. Other minerals found in eastern DRC, such as gold, tungsten, cassiterite and diamonds, are now also only mined using artisanal or small-scale methods, although an industrial niobium mine operated in North Kivu in an area occupied by anti-government forces from 2000 to 2004.

The prevalence of artisanal mining methods in eastern DRC in the 2000s and 2010s is not representative of the DRC mining sector as a whole. Industrial mines were established in the DRC during the colonial era, especially in Katanga (for copper and cobalt) and in East and West Kasai (for diamonds). However, the efficiency and profitability of industrial mines declined markedly in the 1980s and 1990s.

There are an estimated 750,000 to 2,000,000 artisanal miners in the DRC. The total number of people (miners and their families) dependent on artisanal mining and small-scale mining of minerals for their livelihood could be as high as 10 million people or 16 per cent of the total population.[1] Of these, perhaps 300,000 were employed in some way in coltan production and trade in 2009. Most of these people, especially traders, are likely to handle other minerals as well.

Coltan is mined using artisanal methods in the DRC for the reasons set out earlier: high demand, attractive prices, physical amenability of coltan ore to artisanal production and weak property rights. The presence of artisanal mining is also connected to the fact that large-scale modern mining of tantalite never had a chance to become established before the onset of violence and political instability. After the early 1990s, firms that might otherwise have sought to exploit coltan by modern methods were deterred from doing so. It is likely that some tantalite deposits in the DRC are sufficiently large to interest transnational mining corporations willing to invest in industrial methods. However, until their property rights are secure, their staff are safe and they have the infrastructure to operate the mine and move ore to processing sites, such firms will not invest.

Natural resources that can be mined using artisanal and small-scale methods (including those having a relative high value to weight ratio, especially gold and diamonds, but also coltan and tungsten) are found throughout eastern DRC, which is also where conflict and rebel forces are concentrated. The reasons for violence in this region are complex, but the presence of natural resources that can be profitably extracted using artisanal methods has enabled armed groups to profit from their exploitation in a way that would not have been possible if they controlled territory where coal, iron ore and copper mines are found.

High prices for tantalite and the collapse of other economic sectors have made Congolese eager to exploit coltan. Lacking investment capital that would allow the purchase of complex machinery, Congolese have had to use artisanal mining methods or establish small-scale mining operations that rely on basic machinery. Lacking the infrastructure and protection that would normally be provided by a state, Congolese have also resorted to small-scale transport, resulting in the network of small traders or *négociants* in Congolese French (because they negotiate prices) that is so distinctive of the coltan trade. Large consignments of ore cannot be transported effectively due to terrible roads (although air transport is occasionally an option), and their large size makes them vulnerable to extortion. The answer has been for multiple porters and *négociants* to become involved, creating risks relating to frequent exchange, but a reduced risk that an entire large consignment will be lost.

During the Congo War, continuing demand for coltan reinforced the importance of economic networks, whether they were ones that had survived the war or more recent ones created by the Rwandan army and its main Congolese ally, the Rassemblement Congolais pour la Démocratie–Goma (RCD-Goma). Using these networks would have been cheaper and easier than setting up a completely new trading system that bypassed armed groups. At the end of the 2000s many *comptoirs* (minerals trading firms) continued to have business partners in Rwanda, which is a key staging post for coltan on its way to the international market.

Organization of coltan production
Artisanal and small-scale mines for coltan and other minerals can look and sound chaotic: a brown, red or yellow pockmarked landscape denuded of vegetation; crude wooden

ladders descending into pits to allow access; hoses snaking from water sources into pits; and miners in various states of undress, because of physical exertion, standing up to their waists in water, wielding picks and shovels. There is a lot of noise: from generators, the sounds of sluices and from miners hollering at each other. Following rain, pits become flooded and mines become even more dangerous as pit walls and hillsides become less stable. Artisanal mining also occurs in abandoned industrial mines or in networks of manually dug shafts, with dark entrances leading to underground chambers where miners may remain for days.

Despite such appearances, coltan production is organized: claims are owned (through occupation, if not according to the law); miners work in teams of three to six for the owner of the claim; there is a formula governing how ore that is produced is divided among stakeholders; and the entire site is usually controlled by a single authority, often an armed group.

Under DRC law, subsoil rights – mineral deposits underground – belong in the first instance to the state (similar laws exist in many countries). If a mining company or anyone else wants to mine those minerals, they have to buy the right to do so from the state. Artisanal and small-scale miners have no property rights under DRC mining laws; although since 2003 they are permitted to mine if they pay a fee to the state. In 2008 the government zoned areas of North Kivu solely for exploitation by artisanal miners. In practice, because the rule of law is weak, landowners may directly sell to others the right to mine on their land, or they dig up the ground them-selves. Landowners with commercial title to property are also often absent – living in more comfortable, distant towns – and employ managers to run their farms. There are reports of absentee landowners' managers informally selling mining rights without permission. Armed groups and powerful indi-viduals (such as politicians or militia leaders) may also seize control of land to which they do not have legal or customary title, and either directly organize mining themselves or sell the rights to others.

At the top of the mine hierarchy is a single individual, such as a mine manager representing the mine owner or the most senior member of the armed group controlling the mine. In some mines these individuals are known as *chef de colline*, literally 'chief of the hill'. These individuals wield great power over the mine, including who gets work, who is allowed to enter the mine (miners, traders, prostitutes and other visitors) and what prices miners receive for coltan. They are assisted by a team with clear divisions of labour and responsibilities. *Chefs de colline* in coltan mines in Kahuzi Biega National Park had teams of at least ten individuals assisting them: a president, director general, secretary general, a *chef de groupe* (who collects fees from workers), a union representative, a *chef de camp* (in charge of non-mining camp management), a *commandant du camp* (in charge of security, who represents the local police or armed group), a *chef de chantier* (in charge of site infrastructure and logistics) and a number of individuals in charge of prospecting for more coltan deposits. The Bisie mine had a dedicated *'taxateur'* whose role it was to collect taxes.

In the DRC the person who does the hard dirty work of digging up ore is a digger or *creuseur* in Congolese French. There are a small number of reports of Congolese families moving as a unit to coltan mines, but most miners leave their families behind. Some *creuseurs* are seasonal workers, moving to mines to supplement income during lulls in the agricultural season. In some mines, *creuseurs* are both miners and combatants. As in mines around the world, coltan mining communities in the DRC are overwhelmingly young and male – a factor that contributes to the extensive sex industry at mine sites. Of the Bisie mine's estimated 13,000 inhabitants, 65 per cent were men, 30 per cent were women and 5 per cent were children.

Not all miners at coltan deposits are there through their own free will. In 2001 at the Numbi coltan mine in South Kivu, the occupying Rwandan army brought an estimated 1,500 prisoners from Rwanda to mine coltan, in return for a reduction in sentence and a small amount of cash.

Child labour is a common phenomenon in the Congo, and tens of thousands of children work as labourers in artisanal and small-scale mines. The manager of one coltan mine openly admitted that he accepted children over the age of twelve. The impact of coltan fever on the Congolese popular imagination in the late 1990s and early 2000s can partly be gauged by the presence of child labour. Child labourers interviewed by the Pole Institute, one of the few organizations to undertake first-hand field research in coltan mining communities, spoke about why they went to work in coltan mines. In most cases they dropped out of school because they believed coltan mining offered a better future than an education. Halera (aged sixteen) and Safari (aged seventeen), two former schoolchildren who became coltan miners commented: 'Sometimes we earn 100 dollars, more than our parents ever earned. We buy radios and clothes and God willing we will be able to marry in two years, long enough to save something.'[2] Many children also turn to mining because their families cannot afford school fees. Other child miners have never been to school, but calculate that mining coltan offers better opportunities than agriculture. As Halera and Safari stated, 'Look at some of the [coltan] traders who never went to school; they are better off than the teachers who studied.' A teacher interviewed in 2000 commented: 'We are witnessing the emptying of schools. More than 30 per cent of our children drop out of school to mine coltan. Teachers also leave school to mine coltan. In the schools of Mishavu and Kibabi five to ten per cent of teachers have left.'[3] When asked how children survive in coltan mines, the teacher replied:

Box 2.1   Measuring coltan

Trade in any commodity relies upon agreement between sellers and buyers about how much will be paid for a certain quantity. In markets and mines where scales are available and considered reliable, and buyers and sellers trust each other, scales are generally relied upon. In Africa, metric measures are used, so commodities are weighed in grams and kilograms.

In artisanal and small-scale coltan mines, sellers often do not trust buyers and suspect that buyers will try to cheat them by using weights that give a false measure in favour of the buyer. Similarly, buyers do not trust sellers and suspect that a quantity of coltan may include worthless stones that boost the weight. One former trader remarked that sellers 'put nice coltan on top, and nice cassiterite underneath. As both ores look similar it is difficult to tell the difference unless one is an expert.'[4] Even if buyers are sure what they are buying is genuine tantalite ore, reliable scales can be difficult to find. As a consequence, buyers and sellers resort to standard size vessels, such as those used for widely available consumer products, and a price evolves for that 'measure'.

Measures used in coltan mines in eastern DRC have included a commonly available plastic bottle, the tin used for 'le gosse' label condensed milk (which held about 280 grams of coltan) and a dessertspoon. Thus coltan was traded by the dessertspoon or 'le gosse'. What measure is used does not matter, as long as the unit of measure is widely accepted.

> Some give themselves up to the evils of the mines: under-age debauchery, drug-taking, drunkenness. With the dollars they earn they ignore their parents and especially their teachers. The worst is that they manage to incite other children who stayed at school to follow them.

Artisanal miners in the DRC earn on average between $1 and $3 per day. Coltan miners' income is determined in a number of different ways, depending on the mine, the miner, plot title owner and the availability of cash. Some *creuseurs* work as day labourers and get paid in cash at the end of the day. Others get paid cash per kilogram of coltan they produce. At one mine in 2001, *creuseurs* received $3–5 per kilogram of coltan (unprocessed, tantalite ore). Others receive a share of the coltan produced, and do not receive a wage for their work.

Access to coltan deposits and being able to work at a coltan mine requires paying a fee to a person or group that controls the mine. At one mine in 2000, the fee to mine a plot measuring 6 square metres (67 square feet) was $500 per year, payable to the plot owner. At another mine the fee to mine a plot measuring 18 square metres (200 square feet) was between $300 and $1,500. At a third coltan mine, the fee for a 500 square metres (800 square feet) concession cost $4,000. The variation in fees was probably partly due to the varying tantalum pentoxide content of the deposits. In addition to paying a fee to mine a specific plot, *creuseurs* may also have to pay a fee to work at the mine itself. The weekly fee to work in one mine was two dessertspoons (literally) of coltan, then worth about $7.50; one spoonful went to the armed group in control, and one to the *chef de colline*. In 2009, artisanal miners were permitted to mine if they bought a *carte de creuseur* from the government, which cost $25 and was valid for one year. The cards are valid for only one mining zone, and as *creuseurs* often travel from zone to zone this means they either have to buy a new card or – as happens more frequently – simply mine illegally, which makes them vulnerable to extortion by mine managers, police and armed groups.

Cartes de creuseur are also subordinate to companies that have formal rights to a concession. This is less relevant to *creuseurs* mining coltan because there are no corporations mining it. However, if peace comes to eastern DRC and mining companies return, hundreds of thousands of *creuseurs* will have no rights to continue mining and companies will want them to vacate deposits, unless they are in a dedicated artisanal zone. Given the DRC government's past willingness to violently evict artisanal miners, it is highly likely the government would do so again in eastern DRC. Some *creuseurs* may be able to find work with a modern mining company, but most would probably lack the skills required to extract coltan using modern techniques.

Artisanal and small-scale miners are politically weak. If

threatened by anti-government armed groups the DRC state is not very likely to intervene, even if it could. By contrast, states tend to respond with alacrity to any threat of violence or sabotage that target single natural resource developments producing important flows of tax revenue. In Walikale, North Kivu, coltan miners have organized themselves into collectives precisely in order to advance their economic and social interests, but these remain politically weak and do not appear to have had much of an impact.

The way labour is organized in artisanal and small-scale coltan mines elsewhere in Africa is different to the DRC. In Nigeria, artisanal tantalite mining is organized by small groups, often based around family structures. Tantalite mining in Ethiopia uses small-scale methods, but is controlled by a state-owned mining company. As in the DRC, it is possible that there is stratification of ownership and control along ethnic lines. In any case, stronger rule of law in Nigeria, Ethiopia and Rwanda is likely to make mining rights more certain and means that coltan mining in these countries is less prone to violence, which in turn is likely to result in lower turnover of both labour and exchange of rights to plots.

The organization of labour in modern tantalite mines outside the DRC is also different from the way labour is organized at artisanal mines in the DRC. At modern mines, labour is often unionized, relatively well-paid and skilled at operating machinery. In remote areas in wealthy countries, workers often fly in to the mine to do a two-week shift, stay in dormitories where everything is provided by the companies, and then fly out to their permanent home at the end of the two weeks. These mines can still be rough places, but there is a police presence. There may be disputes with indigenous communities over ownership of land and the use of revenue from the mine, but these disputes rarely interfere with production and are usually fought in big city courts far from the mine site.

Illegal coltan production
One of the manifestations of weak state capacity and conflict in the DRC is mining in areas where it is not permitted under law. The DRC national parks authority had been one of the most effective and capable state institutions in the country. It received international funding for research and for protection of conservation areas, parks staff were relatively well paid and well trained, and income from tourists visiting national parks to see gorillas was reinvested into the parks. The national park system has, however, been severely affected by conflict and reduced state capacity. The job of park rangers has become particularly difficult. Rangers have been attacked and killed by armed groups; many go for months without pay, are isolated and receive supplies irregularly, and are resented by people living illegally in the parks. When staff take action against illegal settlers, such as by clearing settlements and confiscating livestock, they face intimidation and reprisals.

Civilians have fled into national parks because the forest allows them to hide from marauding armed groups. People have also moved into parks to exploit their resources. Logging, mining, hunting, fishing and clearing of land for agriculture – some of it organized by armed groups – has occurred in national parks across the DRC. Mining is occurring in two world heritage sites – Kahuzi Biéga National Park and the Okapi Wildlife Reserve – as well as in the South Masisi Protected Nature Reserve, in North Kivu.

Kahuzi Biéga National Park is a 6,000 square kilometre park overlapping North Kivu and South Kivu. It is renowned for its extensive tropical forest that spans lowland areas and high altitude volcanic areas rising to 3,300 metres. The park was primarily created to protect the eastern lowland gorilla, about 85 per cent of which lived in the park and surrounding forests. The park has about 10,000 plant species, 220 bird species and 130 species of mammal, including bush elephants and chimpanzees. During the Congo War, the Rwandan army, RCD-Goma, Rwandan Hutu rebel groups and Mai Mai forces operated within the park and controlled different areas, including routes to coltan mines. In the early 2000s staff controlled no more 10 per cent of the park, although by 2006 they had regained control of 50 per cent. Most of the mining is for cassiterite, but there are also tungsten, cobalt and gold deposits. There is conflicting data on whether coltan continues to be mined within the park. One report from 2003 stated that coltan was mined at twenty-five sites within the park, but research from 2009 suggested that coltan was only mined on the park's northern edge at Ibanga – a site controlled by FDLR (Forces Démocratiques pour la Libération du Rwanda) and Mai Mai forces. At various times during the early 2000s there were between 8,000 and 15,000 miners operating in the national park and an estimated ninety-nine mines in total. Under both DRC law and ordinances passed by the RCD-Goma during the Congo War, all mining in the park was illegal.

The impact on the park of mining for coltan and other minerals was documented by a Congolese investigator working undercover in 2001 for the Institut Congolais pour la Conservation de la Nature and the Dian Fossey Gorilla Fund. To understand the impact of mining on fauna, miners were asked what bushmeat (meat from wildlife) they ate. They reported that in 1999 they had eaten elephant, gorilla, chimpanzee, buffalo and antelope, but by March 2001 they were eating tortoises, birds, small antelope and monkeys. Larger species had been exhausted by hunting. Dietary evidence of the impact of mining on wildlife was corroborated by wildlife surveys. About 350 elephant families lived in the park in the mid-1990s, but only two remained by 2001. In the mid-1990s, 8,000 gorillas lived in the park, but by 2010 the gorilla population had declined by about two-thirds.

The Okapi Wildlife Reserve is a 13,700 square kilometre park in the Ituri forest in Orientale province. It was established to protect okapi, a giraffe-like antelope found only in the DRC, and its tropical lowland habitat in the Congo River basin. Other notable species include bush elephants, monkeys and birds. During the Congo War the reserve was controlled by the Mouvement pour la Libération du Congo and then the Front de Libération du Congo. In the early 2000s, there were an estimated twenty coltan, gold and diamond mines and 4,000 miners in the reserve. The impact of mining in the reserve is unclear. Parks officials and customary chiefs reported in 2001 that wildlife near mines in the park had been wiped out. Another report found that the Front de Libération du Congo had taken steps to remove coltan miners from the park in order to protect the elephant population. However, given that large quantities of ivory were being traded within the area and a Ugandan colonel was caught trying to smuggle almost a tonne of elephant tusks, it is possible that the Front de Libération du Congo tried to remove coltan miners from the reserve so its forces and Ugandan allies could monopolize both the ivory trade and coltan mining.

Illegal production of coltan and other minerals has also severely affected agricultural land. Ownership of agricultural land, and the right to buy land, has been in dispute since colonial times in eastern DRC. Belgians seized land from Congolese, who were already in conflict over grazing rights, then President Mobutu seized ranches from European settlers in the early 1970s and handed them out to politically favoured individuals and groups, including to prominent individuals from Kinyarwanda-speaking Banyamulenge and Banyarwanda communities. Property rights were thrown into doubt with citizenship changes that affected Banyamulenge and Banyarwanda, a weakening of the DRC legal system and state institutions, and then in the 1990s agricultural lands were occupied by, among others, refugees, displaced people, miners and armed groups.

Miners occupying agricultural lands dug up coltan – and any other minerals – wherever it was found. In some cases grazing land was leased out by farm managers to miners without the knowledge of the landowner, bringing the rancher into conflict with his manager and the miners. In other cases, agricultural land was simply invaded and the owners were unable to stop the new occupants. Once grazing land has been mined it is so pitted and denuded of topsoil that it has no value for agriculture. The Goma-based Pole Institute concluded that:

> the massive destruction of former grazing land is catastrophic. Soil which has been used for unplanned prospection and artisanal coltan mining is no longer usable for agriculture. Entire hills and valleys have been turned into giant craters, turning the landscape of the region into an expanse of naked earth, at the bottom of which flow rivers and streams which were diverted for the requirements of coltan mining.[5]

Landowners face the total devaluation of their asset if they are unable to gain any revenue from renting land to miners. Smallholders or peasants who used their land to grow subsistence crops faced a similar situation if coltan was found on their land.

Environmental impact of coltan production
The removal by miners of vegetation and soil to obtain coltan is the main cause of environmental damage. Because artisanal miners have no geological knowledge or access to geological information, they target those areas where deposits break through the surface of the land. However, artisanal and small-scale technology is inadequate to mine a surface deposit far into the ground, so mines are abandoned when it is no longer technically feasible to use these methods, and the miners

move on to another location. One of the main reasons coltan mining has such an impact on the environment is the large quantity of water needed to separate out coltan from other ore.

The absence of a government capable of addressing the environmental impacts of mining, or enforcing mining safety standards, means that pollution of waterways, land degradation and deforestation are totally unregulated. The new mining code introduced by the government of Joseph Kabila in 2002 stipulates that miners are required to restore topsoil following removal of ore, but this does not happen in artisanal mining areas. Only large transnational corporations are capable of doing this and are vulnerable to government pressure to do so. The environmental effects of artisanal and small-scale mining for coltan and other minerals in the DRC include:[6]

- forest clearance to expose soil for mining;
- cutting of timber to build workers' camps;
- cutting of firewood;
- removing the bark from trees to make panning trays to wash coltan;
- pollution of streams by silt from washing process;
- diversion of streams from their original course;
- cutting lianas to make baskets to carry coltan;
- hunting animals, including for food, ivory and other body parts;
- animals injured after escaping snares;
- disturbance of fauna due to people resident in, and moving through, reserves;
- reduced population of invertebrates and reduced photosynthesis in aquatic plants due to silting of streams;
- reduced fish stocks in lakes and rivers affected by silt pollution;
- erosion, including landslides, of unprotected ground during rains;

- ecological changes due to loss of key species, such as elephants;
- long-term changes in watershed due to rapid run-off in deforested areas.

An ethnic dimension

The people who own the land on which mines are situated, those who manage and control mines, those who trade coltan, and those who actually dig up the coltan, can often be distinguished along ethnic lines. A key reason for this is that ethnic networks may provide security for their members in the absence of a capable state. Curiously, the ethnic dimension of coltan production has received only minimal attention from researchers, even though ethnic-based violence is a widely accepted feature of conflict in the DRC.

In North Kivu, mines are often on land owned by Kinyarwanda-speaking Congolese who trace their ancestry to Rwanda. Kinyarwanda speakers can be either Tutsi or Hutu, but their identity in the DRC has historically been based more on their language (Kinyarwanda) and ethnicity (Rwandan roots) than the socioeconomic divisions that define their Rwandan counterparts. Kinyarwanda-speaking Congolese Tutsis own a lot of land in North Kivu and South Kivu as a result of transactions made during the Mobutu era after they were granted citizenship rights. While Congolese Tutsis own the land on which many mines are situated, they do not always control the mines and are not *creuseurs*. The ethnic group that owned much of the land before it was appropriated by the Belgian colonial state and then sold under commercial title during the Mobutu era is the Hunde, and many mine managers and coltan miners belong to this ethnic group. Hunde have a grievance against Congolese Tutsis because of the transfer of lands in the past, and there is underlying tension that occasionally erupts into violence.

However, the most populous group in North Kivu is neither Hunde nor Congolese Tutsis, but Congolese Hutu. Hunde view Congolese Hutus as immigrants (given their Rwandan origins) and resent them because of their claims for political and traditional representation – which, because of their population size, could overwhelm the claims of other groups. There is thus tension between Hunde and Hutu, between Hunde and Tutsi, and more recently between Tutsi and Hutu. Because of hostility between Hunde and Congolese Hutu, there has also been tension and violence between Hunde miners and Rwandan Hutu rebels who entered the DRC in 1994 following the Rwandan genocide.

The risks faced by coltan miners, therefore, are not just related to safety or disease but include ethnic-based attacks. When asked about the coltan industry, the president of a Hunde subgroup, the Bushenge community, commented on the ethnic dimension:

> At the moment [coltan mining] is a high-risk job especially if you are not a Hutu or a Tutsi. Our young Hunde miners are shot at point-blank range. In the mines, when people talk of armed groups they are mainly Interahamwe who systematically kill the Hunde but leave out the [Congolese] Hutu after they have robbed them. The young Hunde who manage to escape from the Interahamwe are then victims of the Tutsi army on the road.[7]

Other coltan mines in North Kivu were controlled by Mai Mai forces but owned by the Nyanga ethnic group. Nyangas, like Hunde, view Kinyarwanda-speaking Congolese as immigrants and resent their political influence. Because of their grievances against Congolese Hutus, Nyanga have been targeted by Rwandan Interahamwe Hutu forces. Mai Mai forces, which controlled coltan mining in Walikale until recently, are predominately of Tembo, Rega or Hunde ethnicity, and have targeted miners of the Shi ethnicity as well as Congolese Tutsi who they considered to be collaborators with the occupying Rwandan army.

In South Kivu a larger number of ethnic groups are involved in coltan mining. Around the Numbi coltan mine the land is mostly owned by members of the Havu ethnic group, although highland areas were owned by Kinyarwanda-speaking Congolese (mostly Tutsi, but also some Hutu). There are also some owners from the Tembo, Shi and Nande ethnic groups. *Creuseurs* themselves are predominately Shi and Havu, and a minority are Hunde.

One reason ethnic networks are likely to have arisen in the mining industry is because they provide some security of exchange and trade through reputation-based coalitions. That is, members of the same ethnic group are more likely to trust each other. A dishonest trader of the same ethnic group is also more likely to be sanctioned through the informal mechanisms of the coalition. The benefits of participating in ethnic networks in conditions of violence and uncertainty result in the reinforcement and expansion of these networks, making ethnicity more and more important over the lifetime of the operation of a coltan mine. The sharing of a common language by members of the same ethnic group may also contribute to collaboration along ethnic lines, although this is notably *not* the case for Congolese Tutsi, Congolese Hutu, Rwandan Hutu rebel groups and Rwandan army troops that occupied the DRC during the Congo War, all of whom speak Kinyarwanda.

Given that mining appears to require stability for optimal exploitation, why is there violence around mine sites? Armed groups use violence against civilians for many different reasons, including to demonstrate their power, to undermine existing authorities, to obtain compliance, and for revenge. The reason armed groups controlling mine sites are able to get away with using violence for these purposes is that labour is cheap

and miners lack collective bargaining power. Unlike industrial mines which require a skilled, healthy and disciplined
(and therefore more expensive and often unionized) labour
force, artisanal mines require workers with few skills beyond
physical strength. As there is a ready supply of such workers in eastern DRC, armed groups are able to target workers
as long as violence does not interfere with the mine workers'
collective demands for food, housing and beer. In a unionized
industrial mine the assault or murder of a worker by mine
authorities would probably cause his colleagues to strike,
protest or sabotage operations. This would interrupt production and reduce profits for those who control the mine. By
contrast, miners in coltan artisanal mines are dispensable
because they are easily replaced and because their colleagues
are less likely to engage in collective action. Artisanal mining
does not require stability in the labour force for optimal production, and this makes violence a viable option by armed
groups controlling mines.

Organization of markets

The movement of coltan from a mine onto the domestic
market and eventually the global market is controlled by *négociants* and other traders. Coltan may transfer through half a
dozen *négociants* before it is purchased by the firm that actually
exports it to processors in Europe, North America and Asia.
Frequent trade, along with artisanal production methods, are
the key features distinguishing the coltan supply chain from
the broader global tantalite supply chain.

Some *négociants* visit mines and buy ore directly but, as
most coltan mines are isolated, coltan is usually carried out of
mines on the heads and backs of porters:

> The coltan grit is bagged in small nylon bags sewn from
> larger food sacks . . . When the bags are full they may weigh

> from 15 kg to 50 kg (33–110 lb) according to the strength of
> the carrier . . . The bags are sewn shut and transported on
> the back in a *makako* – a sort of basket-rucksack made from
> forest lianas.[8]

Porters are particularly vulnerable at this point in the supply
chain. They carry the *makako* through jungle and farmland,
until they reach roads and villages where *négociants* operate.
Like miners, porters are often paid in coltan by *négociants*.

Coltan then changes hands repeatedly until it is bought by
a *comptoir* or minerals trading firm. Unlike *négociants*, *comptoirs* have an established shopfront in a major town. Goma,
Bukavu and Bunia are the major locations for *comptoirs* in
eastern DRC. Unlike coltan miners, *comptoirs* are more organized, more politically powerful and have engaged in collective
action. Their political power is not surprising given that they
are a major source of revenue for the authorities. The fact
that authorities need their money gives *comptoirs* bargaining
power. Their relatively small number, in contrast to the far
more numerous miners and *négociants*, also makes collective
action easier. In mid-2008 *comptoirs* in Goma withheld cassiterite ore from export to protest high export taxes imposed by
provincial authorities. From July 2008 to August 2008 – the
month of protest – export volumes dropped 90 per cent, only
to bounce back in September once the protest was over. The
government decreased taxes in response, so the protest was a
success, but *comptoirs* also diverted consignments to Bukavu
and exported them through that city instead. There was also
anecdotal evidence that smuggling increased.[9] It may have
been the diverting of consignments to Bukavu that persuaded
the North Kivu authorities to reduce taxes, because rerouting
of trade demonstrated that North Kivu could lose the trade
and export taxes to the South Kivu provincial government if
Bukavu became the preferred transport route.

Illicit commerce and smuggling in the DRC is common

and is facilitated by geography and weak state capacity. Coltan mines are relatively close to border areas, which are unfenced and unpatrolled and pass through forested and mountainous areas. Remote airstrips are occasionally used to export ore because of the poor state of roads, the ease of avoiding official-dom and the lack of regulation around aviation. Light aircraft fly in with commodities sought after by Congolese (beer, food, clothing and consumer goods), load up with coltan and other ore, then fly out to Goma and Bukavu or direct to Kigali. During the Congo War some coltan was also transported directly through military airports in Uganda controlled by the Ugandan army.

The organization of coltan production and trade within the DRC is atypical and not representative of the way pro-duction and trade of tantalite are organized elsewhere. Many established minerals trading companies that normally deal in coltan are averse to establishing offices in the DRC because of fears of violence, loss of equipment and loss of investment capital. They therefore rely on intermediary traders – the *négo-ciants* and *comptoirs* that feature so prominently in analyses of the DRC coltan trade – to bring them the ore. The large number of traders who handle coltan before it finally arrives at processing plants in Europe and Asia is distinctive to the DRC. Each time coltan changes hands, money also changes hands and there is uncertainty around the quality and quantity of the ore. Repeated exchanges and the convoluted overland routes that most coltan travels create opportunities for armed groups to extort and tax the trade.

International export of coltan out of the DRC is also fraught with risk. Transportation across frontiers creates logistical and bureaucratic obstacles, and corrupt customs officials demand bribes or seize shipments. Currency and financial services are required to change and transfer money. Congolese who accompany coltan shipments into neighbouring countries find themselves with a valuable commodity without the certainty and support of doing business at home. Foreign businesspeople, too, are ready to cheat Congolese. One trader explained how in 1998 he mortgaged his house for $17,000 to obtain enough capital to buy a 1.3 tonne consignment of coltan from South Kivu. He transported the shipment to Kampala, Uganda, where an English businessman had promised to buy it from him. The trader agreed to let the Englishman store the coltan at a warehouse until its tantalite content had been determined, a process that would take a few weeks as the samples would be sent to London. After the content had been determined the price was to be settled based on the spot market. While the trader was waiting to hear about the tantalite content, the Englishman secretly shipped the coltan to Britain and returned to London himself. The trader was left with neither ore nor payment. After intervention from Interpol and the Ugandan government, the Englishman – who feared arrest if he returned to Uganda without settling his debts – agreed to pay $11,000 for a shipment that was probably worth $27,000 (about $200 per kilogram for ore with 23 per cent concentration of tantalite). It took the trader 10 months to get any return, and he lost his house in the process.[10]

Into the global market
It is at the point of sale between *comptoirs* and international minerals trading firms that coltan enters the global marketplace for natural resources. Many natural resources, including oil, coal and iron ore, are sold on long-term forward contracts by producers to minerals processing firms. A 'forward contract' is a contract between a producer (supplier) and a processing firm (buyer) for the supply of a certain quantity of the commodity at an agreed price for an agreed period into the future. Prices are set through negotiations between producers and processors, both of which use projections of future

demand and supply to determine what they are prepared to offer and pay. This type of market ensures that prices and supplies are relatively predictable, and enables manufacturing firms to better plan and manage costs. However, buyers and sellers generally have no room to negotiate volumes as these are fixed depending on the contract, although there may be some room to renegotiate the price.

Most tantalite is sold on forward contracts, such as tantalite from the Kenitcha mine in Ethiopia and the Wodgina mine in Australia. Tantalite from the Abu Dabbab mine in Egypt (jointly owned by Gippsland Ltd and the Egyptian government), which is forecast to come into production in 2013, has already been locked into a 10-year forward contract to H.C. Starck. Perhaps cognisant of Talison's inability to control costs, Gippsland's contract includes a price escalation clause tied to increases in production costs (designed to enable it to make money even if costs rise), as well as a floor price that will shield it from low prices.[11]

In contrast to the forward contracts through which most of the world's tantalite and some other minerals from the DRC (such as copper) are sold, all coltan is sold on the spot market. 'Spot markets' are open markets without any special contractual arrangements between producers, traders or end users of a commodity. Buying and selling on spot markets is often facilitated by specialist broking firms. A commodity sold on a spot market is paid for in cash and delivered immediately. Prices can fluctuate rapidly, even daily, according to demand. Spot markets are sometimes an actual marketplace, such as the fruit and vegetable markets found throughout the world. For commodities traded globally in large quantities, however, spot markets are a 'virtual' market where sellers and buyers communicate using telephones and the Internet.

There is no central exchange or market where most tantalite is bought and sold, even on the spot market. Tantalite is sold in several metals markets, and some sales are not reported at all. Several specialist minerals broking firms are involved in the spot market, and their role in matching sellers to buyers is an essential one for the efficient functioning of the market, given the lack of a central exchange and readily available information about price and supplies. Daily prices for tantalite are reported in a number of publications, including *Metal Bulletin*, *Ryan's Notes* and *Platts Metals Week*. However, the price varies from source to source depending on the concentration of the ore being sold. Wise buyers and sellers consult all sources of information about prices before agreeing on a price.

Creuseurs and *négociants*, like miners and traders of artisanally produced minerals everywhere, do not receive spot market prices. This is largely because of the remoteness of mines and poor communications, which means they are unable to find out the going price for tantalite on any one day, allowing traders to take advantage of sellers' lack of knowledge. The tantalite content of coltan is also rarely known to *négociants* since they lack the appropriate chemical analysis equipment, making them unwilling to pay spot prices (even if they knew them) because of the risk that the coltan has a low concentration of tantalum pentoxide. Because of their remoteness, poor infrastructure and lack of information, producers of coltan are condemned to using the network of middleman-traders to get their ore to end users. But it is traders who bear the cost and risk of getting the ore to market. It only when coltan reaches *comptoirs* based in big towns with access to international price information that it begins to be traded at a price approaching that set by the global market. Table 2.1 shows the estimated proportion of revenue received by different buyers and sellers along the coltan supply chain.

During the tantalite price boom of 2000, industrial producers on forward contracts regretted those contracts because it meant they were unable to benefit from spectacular price

Table 2.1   Distribution of coltan profits, c.2000

Actor in the supply chain	Distribution of profits	
	in the DRC	Globally
Team of *creuseurs* (6 persons)	17	2
Chief of mine	10	1
Petits négociants	10	1
Gros négociants	13	2
Comptoir	10	1
Taxes to RCD-Goma	7	1
Other licences and fees	22	3
Armed groups	11	1
Revenue in D.R. Congo	*100%*	*12%*
Minerals brokerage firm		14
Processor		27
Capacitor manufacturer		46
Global revenue		*100%*

Source: adapted from Le Billon, Philippe & Hocquard, Christian (2007), Filières industrielles et conflits armés: le cas du tantale dans la region des Grands Lacs. *Écologie & Politique* **34**, 83–92, p. 90. Le Billon and Hocquard based their table on information from Martineau, Patrice (2003), *La route commerciale du coltan: une enquête*. Groupe de recherche sur les activitiés minières en Afrique, Faculté de science politique et de droit, Université de Québec, Montréal; and de Failly, Didier (2001), Coltan: pour comprendre. In: *L'annuaire des Grands Lacs*, L'Harmattan, Paris, pp. 279–306.

rises on the spot market – up to ten times what they received through forward contracts. Of course, the large amount of tantalum supply locked into forward contracts contributed to the price rise in the first place! *Comptoirs*, on the other hand, benefited tremendously because buyers were forced to turn to coltan sold on the international spot market because other sources were tied up in long-term contracts. DRC sources of tantalite were the biggest suppliers to the spot market.

Notwithstanding the windfall profits made by some coltan

producers and traders during the price boom, most producers and traders would benefit from signing forward contracts
because these would guarantee prices. But no processing firm
in its right mind would ever sign a forward contract with a
supplier of coltan sourced from the DRC, and this is a major
impediment to stable development of the minerals sector in
eastern DRC. Political instability, violence, poor infrastructure and artisanal and small-scale production methods make
production and supply of coltan unpredictable, and predict-
ability is essential for being able to forge long-term forward
contracts.

Processing: coltan and tantalite converge
After being sold at a price set by the international spot market,
coltan enters the third step in the tantalum supply chain:
processing. It is at processing plants in Asia, Europe and
North America that coltan can converge with tantalite from
other countries. At these plants tantalite is chemically transformed (reduced) to tantalum metal and then processed into
market-usable forms, such as foil, sheet, plate, wire and pow-
ders. About two-thirds of tantalum is processed into powder
products, as this is the form required by manufacturers of
capacitors who are major end users of tantalum. Figure 2.1
shows the seven steps of the supply chain relevant to tantalum
in a mobile phone, through a hypothetical scenario *c*.2007,
including where the coltan and tantalum supply chains could
come together.

A number of companies process tantalite into tantalum,
but three dominate the market: H.C. Starck (headquartered
in Germany), Cabot (United States) and Ningxia Orient
Tantalum Industry (owned by the Ningxia Non-Ferrous
Metals Smeltery, China). Together they consume 70–80 per
cent of the world's supply of tantalite, and also produce about
80 per cent of the world's tantalum powder – H.C. Starck and

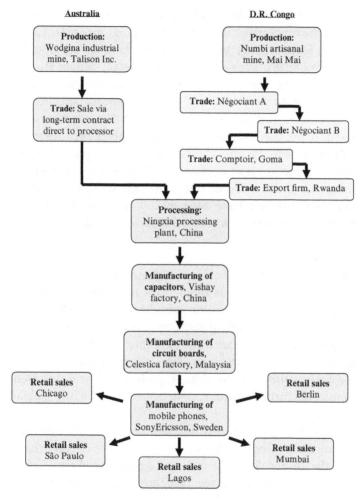

Figure 2.1 Hypothetical supply chain convergence, c.2007.

Cabot produce about 30 per cent each and Ningxia 20 per cent. These three firms have historically set the price for tantalite through their negotiations with industrial producers. They have also vertically integrated their businesses, processing

raw tantalite into tantalum metal and then using this to manufacture components used in complex end products. Cabot
is even further integrated – it owns the Lake Bernic tantalite
mine in Canada. Kazatomprom (Kazakhstan), Mitsui Mining
and Smelting (Japan), Malaysia Smelting Corporation and
some other smaller firms are also able to process tantalite.

Processors of tantalite can be confident that some consignments of ore they receive do not contain coltan, because some
shipments will clearly originate in, for example, Ethiopia,
Mozambique or Brazil. However, if processors choose to
mix consignments and do not inform the downstream manufacturer that they have done this, it is impossible for the
manufacturer to know whether this has occurred.

Manufacturing

Tantalum products are sold to companies that make materials
and components, such as alloys, glass and, most importantly,
capacitors. Many companies manufacture components and
final products containing tantalum. There are three major
stages in the manufacturing of digital electronic devices:
manufacturing of capacitors, then circuitboards, then final
assembly of the end product itself. (Note, not all firms involved
in manufacturing electronic devices also manufacture mobile
phones.)

Capacitors used in electronic devices are made by about
twenty companies from China, Germany, Japan, Great Britain
and the United States in factories across East Asia, Europe
and North America. Large capacitor manufacturing firms
include AVX, Kemet, NEC, Samsung and Vishay. Capacitor
manufacturing is the fourth step in the tantalum supply chain
relevant to mobile phones.

Five companies dominate circuitboard manufacturing (the fifth step in the supply chain for mobile phones):
Flextronics (headquartered in Singapore), Celestica (Canada),

Sanmina-SCI (United States), ASUStek (Taiwan) and Jabil (United States). Like other aspects of the tantalum supply chain, circuitboard manufacturing is an international business involving transnational corporations with factories and offices in several countries. For example, ASUStek is headquartered in Taiwan, listed on the London and Taiwan Stock Exchanges, and has factories in China, Mexico and the Czech Republic.

The next step in the tantalum supply chain involves the manufacture of final end products (step six in the supply chain for mobile phones). It is at this stage that some of the world's biggest brand name corporations for electronic digital devices become involved, such as Acer, Apple Inc., Canon, Dell, Ericsson, Fujitsu Siemens, General Electric, Hitachi, Matsushita, Mitsubishi, Motorola, NEC, Nokia, Philips, Samsung, Sony Ericsson, Texas Instruments and Toshiba. These firms are from Japan, the United States and Europe (Finland, Sweden, Germany and the Netherlands), with some from South Korea and Taiwan. Firms from China are increasingly important as manufacturers of electronic devices and therefore consumers of tantalum capacitors. The Chinese firm, ZTE, is now the sixth largest manufacturer of telephone handsets.

Alloys containing tantalum are made by about twenty-five companies in Australia, Brazil, China, France, India, Italy, Russia, Great Britain and the United States. Alloys are used in chemical process equipment, jet engines, medical instruments, missile parts and nuclear reactors. Companies that use tantalum alloys include Boston Scientific (medical devices), General Electric, Johnson & Johnson, Mitsubishi Heavy Industries, Plansee (an Austrian medical devices manufacturer), Pratt & Witney, Rolls Royce and Ultramet (a Californian medical devices manufacturer).

Research into mineral supply chains, including coltan supply chains, found that consumer electronics corporations are generally not aware of the location and conditions under which the minerals that are used in electronic products are mined.[12] MakeITfair, a European NGO, sent questions to twenty-three electronics corporations about their supply chains. Of the twelve firms that responded, eight said they did not know where the metals included in their products were mined. Only one, Sony Ericsson, said it knew where its metals were from, but only for one metal: tantalum – a direct result of activist initiatives to get transnational corporations to pay attention to the origin of tantalum. There may be a 'head-in-the-sand' element to these firms' lack of knowledge about the origins of the tantalum in their components, but as evidenced by this chapter, tantalum supply chains can be complicated to trace, complicated to control and complicated to audit.

Conclusion

Most tantalite has been, and is, sold on long-term forward contracts. However, a significant minor proportion of global tantalite production – including all coltan – is sold on the international spot market. No single spot market, such as the London Metal Exchange, dominates trade, making it more expensive and difficult to obtain good information and increasing the basic cost of doing business.

Globally, tantalum is produced in dual ways: regulated capital-intensive extraction, especially in rich countries, and small-scale extraction in poorer countries including the DRC. The production and trade of coltan in the DRC involves a combination of factors specific to that country. Weak state institutions are incapable of enforcing property rights or contracts. Production of coltan is carried out by low-paid artisanal and small-scale miners in isolated mines. Trade involves multiple parties, transportation in the initial stages from the mine to *négociant* over land-based routes, and exchange occurs in the absence of written contracts. The control of different mining functions by specific ethnic groups has created tensions that have fuelled local-level violence, as well as the larger regional conflict that engulfed eastern DRC during the Congo War. While ethnicity is a factor in conflict, however, it is not simply because of ancient tribal rivalries. Rather, it is because ethnicity is directly linked to economic interests related to land and minerals.

These features facilitate entry and participation in the coltan supply chain by armed groups because production can be easily controlled through violence. By contrast, capital-intensive industrial production would screech to a halt if mine managers used violence against labour. Opportunities are also created for extortion in the frequent exchange and transportation of coltan. However, to what extent can coltan be singled out as a cause of violence and source of profits for armed groups? The next chapter focuses on the relationship between coltan and conflict, and analyses the social impact of armed groups involved in the coltan trade.

3

Coltan and conflict

Tantalum has received attention from activists, scholars and NGOs as a direct result of the perceived role of coltan in exacerbating violence in the DRC. Natural resources have had a role in many conflicts, and the fact that there is an economic dimension to violence in the DRC that involves natural resources is not unusual in itself. However, following the end of the Cold War in 1989, natural resources seemed to become more prominent in conflicts, and war in the Congo seemed to be an exemplar of a new type of conflict where belligerents had no other objective than personal enrichment.

Conflict is not new to the Congo. Since at least the mid-nineteenth century, Congolese have suffered at the hands of Arab slave-traders and from state-sponsored violence inflicted by King Leopold II's private mercenary army (the *Force Publique*), Belgian colonial agents, and troops in the feared Presidential Special Division of President Mobutu (in power from 1965 to 1997). However, since the collapse of communist regimes in Eastern Europe in 1989, violence and conflict has intensified. The DRC has experienced five overlapping but conceptually distinct waves of violence: tension over citizenship rights that evolved into violence, spillover effects of the Rwandan genocide, the military campaign against President Mobutu, the international Congo War that drew in half a dozen neighbouring governments, and ongoing local violence in the east. It was during the Congo War phase that natural resources rose to prominence as a factor that seemed to motivate armed groups more than any other factor, such as political or strategic considerations.

A common theme in all these phases of conflict is the devastating impact of violence on civilians, especially in rural communities. Armed groups have engaged in widespread sexual violence, kidnapped men, women and children, stolen livestock and possessions, burned houses, and made agricultural work in isolated fields very dangerous. Because of the scale of sexual violence, the Congo has repeatedly been called 'the worst place on earth to be a woman'. Health and education systems are shattered.

A second common theme is the weakness of the state. Unlike the United States, Britain, China or South Africa, for example, the DRC government's military and police are not pre-eminent armed forces capable of subduing anti-government groups willing to use violence. They do not have a *monopoly on violence* within DRC territory. The state's weakness made it easy for the armies of Uganda and Rwanda to invade the DRC, and created opportunities for domestic armed groups, especially anti-government forces aided and abetted by Uganda and Rwanda.

A third common theme is that the interests and strategies of armed groups have evolved. At various times, natural resources have been the most important motivating factor for some armed groups. However, armed groups have always been motivated by a combination of factors, not only natural resources. Coltan has generated profits for some armed groups, especially in 2000, and continues to do so albeit to a lesser extent. However, its significance as a factor in violence has fluctuated over time, from location to location, and in general the significance of coltan has been exaggerated. Coltan is not the most important, or even a major, cause of violence.

Natural resources and conflict

Natural resources have been sought after and fought over for centuries. Of course, wars have also been fought for other reasons, such as succession within royal families, control of trade routes, for religious freedom, for religious dominance, and for independence. After the Second World War, however, conflict seemed to be less about economics and more about ideology and politics. Cold War competition between Western countries and the Soviet Union and its allies and, related to this, wars and campaigns for independence in Africa and Asia from the 1940s to the 1970s, reinforced the apparent dominant role of ideology and politics in conflicts. Wars in China, Korea, Vietnam, Afghanistan, Algeria, Mozambique, Rhodesia/Zimbabwe, Cuba, Nicaragua and El Salvador, for example, were certainly about control of land (a natural resource), but they were also anti-colonial and/or about which political and economic system should dominate the country.

The nature of war seemed to change following the end of the Cold War and the demise of the Soviet Union in 1991. Conflicts that originated in the Cold War era appeared to become bereft of ideological or political content and mainly focused on retaining or expanding control of natural resources. Examples include Angola (oil and diamonds) and Colombia (cocaine and emeralds). There were also new civil wars that seemed to revolve primarily around control of mineral resources, including in Sierra Leone (diamonds), Liberia (diamonds) and the Congo (timber, diamonds, gold, copper, tin, tungsten, cobalt and coltan). A major reason for the prominent role of natural resources in conflicts of the 1990s was that the belligerents had lost their Cold War patrons. Governments and armed groups wanting to wage war had to raise their own funds, and consequently developed economic interests and strategies that were previously not required.

Another reason natural resources became a factor in these conflicts was that corruption and weak governance facilitated the entrance of criminal and opportunistic actors with few political agendas, but an interest in extracting profits from any available sources.

But why did natural resources, as opposed to other industries or economic activities, become so important? Exploitation and trade of natural resources offer many advantages to armed groups engaged in conflict. Unlike manufacturing or large-scale commercial agriculture, primary commodity production does not 'require complex information and transactions, or a steady stream of specialized inputs. Primary commodity producers are therefore able to survive predatory taxation, unlike entrepreneurs in other sectors.'[1] The endurance of primary commodity production during conflict is attractive to armed groups seeking ongoing sources of revenue that require few investments and inputs, as opposed to short-term sources of revenue such as looting. Primary commodities such as minerals and timber are also 'generic rather than branded products, and so their origin is much more difficult to determine'.[2] The ability of businesses to hide or disguise the origin of generic products bought from armed groups in conflict zones is particularly advantageous to non-state and illegal organizations, such as rebel forces.

Primary commodity production, even in the context of war, is also facilitated by private companies. In the case of mining companies whose profits are dependent upon operations in areas that become conflict zones, they are likely to try to continue operations if at all possible. This is because minerals are site-specific and mines can be costly to develop, making them 'condemned to developing resources where they can find them and where their investments have been made, whether that is in a war zone or not'.[3] Congolese civilians involved in mining are even more likely to be 'condemned' to remain in mining, because they have limited ability to migrate out of the region, transfer capital, retrain, or find a job in another sector. Private actors, including some mining firms, will remain in conflict zones, and occasionally even commence investment, despite the risks. The result can be an ongoing process of negotiation between mining companies and armed groups, whereby companies produce revenue for the armed group in return for protection.

Not all natural resources are equally attractive to armed groups. Large-scale industrial mines and oil installations tend to rely on specialized infrastructure and expatriate staff, making violent intimidation and extortion attractive to rebels. However, actual rebel control of production is not an option as investors would stop investing, expatriate labour would go home and infrastructure would break down. The complex inputs required for large-scale extraction of natural resources, as outlined in chapter 1, make it impossible for any organization other than a mining or oil company to develop and operate industrial-scale mines and oil installations. On the other hand, minerals that can be mined by artisanal methods can be low-tech and labour-intensive, allowing exploitation to occur in remote areas and in poorly regulated business environments.

Revenues from natural resources were not central to the initial interests and strategies of armed groups involved in the Congo War. However, within the first year both pro- and anti-government forces, including foreign governments and their armies, developed economic interests. In some cases these interests became more important than political or strategic interests. In other cases, armed groups' economic interests fluctuated in importance throughout the conflict depending upon military success, developments at the political level (such as peace agreements) and the availability of other sources of funding and support for military campaigns.

Conflict in eastern Congo

The violence and conflict that has engulfed eastern DRC over the past two decades has come in five waves that overlap, but are conceptually distinct. The region – namely the provinces of North and South Kivu, and parts of Ituri, Maniema and northern Katanga – was already the site of long-standing tensions over land, citizenship and ethnicity. Some, but not all, ethnic tension was between self-proclaimed 'autochthones' or indigenous Congolese, on the one hand, and Kinyarwanda-speaking Banyamulenge and Banyarwanda communities, including both Tutsis and Hutus, who trace their ancestry to Rwanda, on the other. Tensions focused on which ethnic groups had access to land and the right to own it, and which had citizenship and associated rights. The policies of Belgian colonial officials and President Mobutu created and exacerbated these tensions. Access to land or citizenship, or both, was manipulated opportunistically by both regimes in order to favour certain communities. Increasing population pressure and associated competition over scarce land resources were also a contributing factor to communal violence and conflict.

Citizenship, economic opportunities and ethnic competition
The first wave of violence in the early 1990s was sparked by changes to the citizenship laws, but this had its roots in earlier state policies. In 1972, President Mobutu sought to build support and win allies in eastern DRC by giving Congolese of Rwandan descent full citizenship rights, a politically significant step given that they were a demographic majority in some areas. Previously restricted to being traders and tenant farmers, citizenship enabled them to buy land and to become senior public officials. However, President Mobutu's grip on power weakened in 1991 as political opponents, inspired

by democratization and the downfall of dictators in Eastern Europe, began to challenge his authority. Mobutu also lost the financial support of the United States (and the multilateral institutions which it dominated, such as the World Bank and International Monetary Fund) following the demise of the Soviet Union. Seeking to shore up his weakening power by appeasing domestic critics, in 1991 President Mobutu upheld changes to citizenship laws from 1981 that had annulled citizenship for most Banyamulenge and Banyarwanda. To retain citizenship they had to be able to trace their ancestry in the DRC to 1885, when the Congo became the property of King Leopold II of Belgium. Revocation of citizenship drastically reduced these communities' economic opportunities and understandably caused great insecurity. Tension between the Banyamulenge and Banyarwanda communities on the one hand, and autochthonous Congolese who had resented these communities' previous favoured treatment by Mobutu on the other, escalated into low-level communal violence in 1993. Ownership of and access to land for agricultural purposes was a key focus of tensions during this period.

Spillover effects of the Rwandan genocide
The second wave of violence was directly related to the genocide in Rwanda in 1994, when 800,000 Tutsis and moderate Hutus were slaughtered over three months in a campaign largely orchestrated by the Interahamwe, an extremist organization of Rwandan Hutus. The genocide was stopped by the military campaign of the Rwandan Patriotic Front, an organization of expatriate Rwandan Tutsis based in Uganda. About 2 million Rwandans flooded into eastern DRC to escape the Rwandan Patriotic Front's advancing forces, including Interahamwe who committed the genocide, officials from the Hutu Power organization who masterminded it, and ordinary Rwandan Hutus frightened of retribution from Tutsis.

Refugees crowded into sprawling camps in the Congo that could not be controlled by the DRC government, aid agencies or the United Nations, but were tightly controlled by Hutu Power and Interahamwe who commandeered humanitarian aid. The genocide and the influx of refugees into eastern DRC exacerbated existing tensions between Banyamulenge and Banyarwanda communities and autochthones. Autochthones claimed that the refugees were just another Rwandan 'invasion' and were suspicious of burgeoning relationships between the new Tutsi-dominated Rwandan government and Banyamulenge and Banyarwanda communities. The latter two communities felt threatened by armed Rwandan Hutu groups and some communities were also attacked. Banyamulenge and Banyarwanda reacted by seeking and receiving weapons and training from the Rwandan Patriotic Front. Low-level violence and tension over land escalated into systemic violence between communities along ethnic and citizenship lines.

The war against Mobutu, 1996–1997

The third wave of conflict over 1996 and 1997 resulted in the overthrow of President Mobutu, who had been in power since 1965. Mobutu's government and military forces were unable to control Hutu Power and Interahamwe forces, which used refugee camps in eastern DRC to reorganize and launch attacks back into Rwanda, where the new Tutsi-dominated government feared an invasion. The Ugandan government, the Rwandan government's close ally and sponsor, also did not want a radical Hutu government reinstalled in Rwanda and had been plagued by cross-border attacks from militias based in the DRC. President Mobutu had been unwilling, and probably unable, to do anything about these militias. After three decades of manipulating ethnic grievances to gain support and undermine adversaries, Mobutu himself had many domestic opponents. The governments of Rwanda, Uganda and Angola (like Uganda, Angola suffered attacks from forces based in the DRC) decided to fund and organize a military campaign to depose Mobutu.

The DRC was invaded by the Rwandan, Ugandan and Angolan armies, supported by an alliance of Congolese. Congolese troops were initially recruited from disgruntled communities in eastern DRC, especially Kinyarwanda-speaking communities, but the campaign gathered popular support as it swept across the country towards Kinshasa. The anti-Mobutu forces, the Alliance des Forces Démocratique pour la Libération du Congo-Zaire ('the Alliance'), were headed by Laurent-Désiré Kabila, a Congolese from Katanga. After an 8-month campaign, on 17 May 1997 the Alliance entered Kinshasa and Kabila declared himself President.

During the military campaign foreign mining companies courted Alliance leaders; one private investor even lent Laurent-Désiré Kabila a corporate jet. Some mining companies signed agreements with the Alliance that granted them rights to vast tracts of land, even before Alliance forces had reached Kinshasa. The Alliance used the revenue from selling mineral rights that it did not yet own to fund its military campaign. The Alliance's takeover of the government threw into question property and mining rights as old Mobutu allies fled or had their property seized. Publicly listed mining companies, especially those concerned about reputational costs, had eschewed investing in the DRC during the latter years of the Mobutu regime due to political uncertainty and collapsing infrastructure. However, transnational mining companies such as Anglo American and BHP Billiton were aware of the DRC's mineral resources and with the change in regime quickly investigated opportunities for exploration, mining and joint ventures.

The Congo War, 1998–2003
The fourth wave of violence is what became known as the 'Congo War' from 1998 to 2003, although Congo *wars* would be more appropriate, given the multi-level, overlapping conflicts that occurred during this period into the 2000s. The Congo War was a civil and international 'war complex' that by its end in 2003 had resulted in at least a million deaths, the internal displacement of 2 million people and another 350,000 people seeking refuge in neighbouring countries. Its origins are rooted in tensions between President Laurent-Désiré Kabila and his Rwandan and Ugandan allies. Some Congolese government officials came to resent the influence of Banyamulenge and Rwandan advisers; Laurent-Désiré Kabila refused to cooperate with Rwanda in arresting Interahamwe and Hutu Power officials present in eastern DRC who were responsible for the Rwandan genocide; and Congolese officials accused Ugandan officials of profiteering.

Renewed violence broke out in August 1998, when a Rwandan and Banyamulenge force attacked government troops and tried to capture Kinshasa. Foreign governments then launched into the fray in support of either government or anti-government forces. After 10 months of fighting the conflict reached a military stalemate and by mid-1999 the DRC was divided into two roughly equal halves by a front line that stretched from the northwest to the southeast of the country. The eastern half was controlled by three main militias: the Mouvement pour la Libération du Congo (MLC) backed by the Ugandan army, the Rassemblement Congolais pour la Démocratie–Mouvement de Libération (RCD-ML) also backed by the Ugandan army, and the Rassemblement Congolais pour la Démocratie–Goma (RCD-Goma) backed by the Rwandan army. The western half was controlled by government forces, backed by the Zimbabwean and Angolan militaries. Loosely aligned to pro-government forces were

Mai Mai militias who operated across eastern DRC. Mai Mai militias were composed of autochthonous Congolese who proclaimed themselves to be anti-foreigner, 'foreigners' being Rwandans, Ugandans and Congolese Kinyarwanda-speakers.

The motivations of armed groups involved in the Congo War evolved as the conflict progressed. While political and strategic objectives were important at the start of the conflict, all armed groups turned to revenue-raising activities to finance their costly military campaigns. As the conflict wore on, economic interests became a major reason for foreign forces to stay and to continue fighting. Their control of mines, forests, borders and trade routes all created opportunities for profit-making. Having started as a conflict in which political and strategic agendas were paramount, the Congo War became a conflict in which economic agendas became just as important as other agendas, and at times more important than other interests. Table 3.1 illustrates the major military actors that participated in the Congo War and their sources of income from minerals. The DRC government rewarded the Zimbabwean and Angolan governments for their support by granting mining rights to firms owned by these governments or their political elites, as well as permitting mining joint ventures between these firms and DRC state-owned companies. Some Mai Mai militias opportunistically cooperated with anti-government forces especially in order to exploit natural resources, including coltan.

President Laurent-Désiré Kabila was assassinated on 16 January 2001, and replaced by his son, Joseph Kabila. The latter was subsequently elected president in 2006 in the DRC's first democratic elections since 1960. Joseph Kabila's government has not attempted to address the root causes of historical grievances regarding land and citizenship in eastern DRC.

Table 3.1   Major armed groups and sources of mineral revenue, 1998–2003

Armed group	Minerals						
	Coltan	Gold	Tin	Tungsten	Diamonds	Copper	Cobalt
Pro-government forces							
DRC army					✓	✓	✓
Zimbabwean army					✓		
Angolan army					✓		
Mai Mai	✓	✓	✓		✓		
Anti-government forces							
Rwandan army	✓	✓	✓	✓	✓		
Ugandan army	✓	✓	✓	✓	✓		
RCD-Goma	✓	✓	✓	✓	✓		
RCD-ML	✓	✓	✓	✓	✓		
MLC	✓	✓			✓		

Source: table compiled by author using data from United Nations reports.

Ongoing local violence

The fifth wave of violence involves ongoing conflict at the local level. In 2009, six years after the official end of the Congo War, anti-government forces and other militias continue to operate in eastern DRC even though DRC government authorities and its army, the Forces Armées de la République Démocratique du Congo ('DRC army'), returned to the east following the peace agreement and Ugandan and Rwandan forces have withdrawn. The authority and power of government and military officials answering to Kinshasa is largely confined to major towns.

The total number of direct and indirect deaths attributable to conflict since 1998 reached well over one million by 2008, although one estimate put the total at about five million.[4] Over half those deaths have occurred since the formal peace agreement and are largely attributable to ongoing local-level violence. While some refugees and internally displaced people returned home after the end of the Congo War, 1.4 million people remain internally displaced. In August 2008, 500,000 people were newly displaced as a result of renewed violence in North Kivu.

The most significant militias in the late 2000s were the Front Démocratique pour la Liberation du Rwanda (FDLR) and the Congrès National du Peuple (CNDP). The FDLR is a Rwandan Hutu rebel group with about 6,000 combatants, including some former Interahamwe. It receives some support from elements of the DRC army. The CNDP was dominated by Banyamulenge and Banyarwanda Tutsis, linked politically and militarily to Rwanda. A third militia is the Patriotes Résistants Congolais (PARECO). PARECO comprises non-Banyamulenge Congolese who are politically aligned with the DRC government and hostile to the CNDP and Rwanda. A fourth anti-DRC government militia, the Front Populaire pour la Justice au Congo, is active in Ituri province. In addition to these larger militias, there are a myriad of Mai Mai groups that have continually splintered, merged and de-merged.

In August 2008 in North Kivu, there was extensive fighting between the DRC army, FDLR, PARECO and Mai Mai groups on the one hand, and the CNDP on the other. Then in early 2009, in the most significant political development since the end of the Congo War, the DRC and Rwandan governments cooperated to attack and disperse the FDLR and then to demobilize the CNDP. The Kabila government permitted Rwanda to send 4,000 troops into the DRC as part of Operation Umoja Wetu (Swahili for 'Our Unity'). The DRC government, sitting in far-away Kinshasa, had realized it could not defeat the CNDP alone and that the FDLR had become more of a hindrance than a useful ally. The Rwandan government concluded that it had more to gain from cooperating with the DRC government than undermining it by backing the CNDP, because cooperation on strategic issues promised to bring Rwandans new economic opportunities through an expansion of trade and commerce. Umoja Wetu had a limited impact on the FLDR, succeeding in dispersing various FDLR groups rather than defeating them. However, it did bring about the end of the CNDP. The Rwandan government withdrew support for the CNDP and put pressure on it to disarm. The CNDP's leader, Laurent Nkunda, was eventually arrested in Rwanda in January 2009. Elements of the CNDP were subsequently incorporated into the DRC army, but this has been an uneasy integration. Former CNDP officials use their positions in the DRC army to extend and expand their personal business networks – as do members of the DRC army – including the exploitation and trade in minerals. Economic partnerships between former CNDP officials and the Rwandan government also remain in place. Table 3.2 shows the sources of mining revenue for selected armed groups between 2006 and 2008.

There are several causes of local violence: top-down manipulation in the form of military campaigns to achieve regional political ambitions, such as by the Rwandan army, the Ugandan army, the FDLR and the CNDP; retaliation and settling of scores related to events during the Congo War; struggles for political control over communities between new militia-based leaders and returning former leaders (such as customary chiefs); conflict between ethnic groups based on which groups received positions in the post-war government; conflicts over agricultural land; conflicts over natural resources; illiterate and poorly educated men being attracted to armed groups in order to become socially important; and a basic struggle for survival, especially by young men who have missed out on education because of the war or have not been allocated land by family or chiefs.

Table 3.2 Major armed groups and sources of mineral revenue, 2006–2008

Armed group	Mineral							
	Coltan	Gold	Tin	Tungsten	Manganese	Diamonds	Copper	Cobalt
DRC army	✓	✓	✓	✓	✓	✓	✓	✓
Mai Mai	✓	✓	✓			✓		
PARECO	✓	✓	✓					
FDLR	✓	✓	✓	✓				
CNDP	✓	✓	✓		✓			

Source: table compiled by author using data from United Nations reports and the International Peace Information Service, Ghent, Belgium (2009), *Interactive map of militarised mining areas in the Kivus*: www.ipisresaerch.be.

Figure 3.1 shows the major armed groups engaged in conflict in the DRC between 1998 and 2009, and indicates where cooperation occurred between groups.

Peace initiatives

A peace process signed in late 2002 and concluded in April 2003 brought an end to the international-level Congo War involving foreign governments. Under the terms of the agreement, foreign armies withdrew and a transitional government was formed that included representatives from the major anti-government forces and civil society. However, for local communities in eastern DRC there has been little change in their living conditions and basic human security. Anti-government forces and Mai Mai continued to use violence against civilians, to tax commerce and to exploit natural resources, including coltan. The various peace agreements signed by participants in the Congo War referred to economic interests and called for an end to illegal exploitation of natural resources, but there was never any serious attempt by the parties to these agreements to address these issues.

To support and monitor the withdrawal of foreign forces from the DRC and the disarmament of anti-government forces, United Nations member states agreed to establish a mission in the DRC, the Mission de l'Organisation des Nations Unies au Congo (MONUC). MONUC was not supported by all governments, and its mandate was consequently limited to observing ceasefire violations and voluntary disarmament, demobilization, repatriation, resettlement and reintegration programmes for different rebel groups. The limited mandate meant MONUC could defend citizens when they were attacked but could not actively pursue militias. However, the mission and its mandate have continued to evolve. MONUC has grown to 22,000 military personnel and police, and its

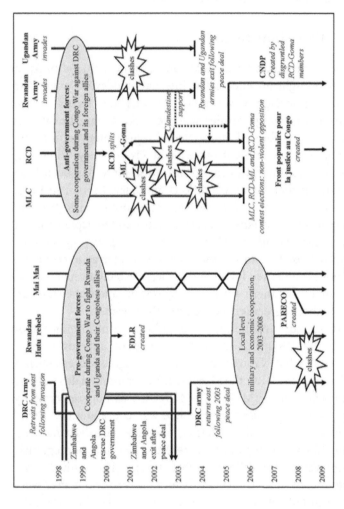

Figure 3.1 Major armed groups' engagement in conflict, 1998–2009.

mandate was strengthened and expanded in 2008 to include interventions focused on curtailing the exploitation of natural resources by illegal armed groups.

MONUC has been heavily criticized by Congolese who see it as an ineffective force that has allowed conflict to continue, as instanced by its apparent inability or unwillingness to stop the CNDP's campaign of violence in North Kivu in late 2008. In 2010 the DRC government announced that it wanted all MONUC personnel to leave by mid-2011, before the next elections are due. MONUC has, however, had some successes. Notwithstanding incursions by Rwandan troops into the DRC since 2003, foreign armies have withdrawn. During the Congo War there was talk of the DRC breaking up into different states but it has remained whole, something most Congolese seem to want. The United Nations' opposition to the DRC breaking up, a principle that shapes MONUC's operations, has probably discouraged fledgling secessionist movements. Thousands of soldiers and militia members have handed over their weapons, demobilized and returned home as a result of United Nations programmes. With the expansion of MONUC's mandate it is more likely to satisfy civilian demands for protection and for active intervention against the economic activities of illegal armed groups. The United Nations also played a key role in organizing democratic elections in 2006, which are widely considered to have been free and fair.

There have also been peace initiatives at the local level. These were mostly implemented and funded by NGOs; some received funding from the diplomatic missions of Sweden, Belgium and the United Kingdom. Local peacebuilding initiatives have included promoting dialogue between rival armed groups, public education about the benefits of grass-roots collaboration, football matches between opposing villages, micro-credit schemes for women from rival clans and groups, and resolution of land conflicts. Between 2002 and

2004 MONUC also had a peacebuilding programme targeting local-level violence, but this ceased when the individuals championing the programme within MONUC left the organization. Local peace initiatives have not gained more support from the international peacebuilding community – diplomats, UN employees and donor staff – for two main reasons. First, peacebuilders do not believe that resolving conflict at the local level is their responsibility (they think they should focus on conflict involving large, organized, rebel groups and governments). Second, they lack the skills required to build peace at the local level, such as out in the countryside dealing with village chiefs, local youth and market women.

Some collaboration around coltan mines that resembles local-level peacebuilding is deceptive. In South Kivu local administrative officials and civil society groups collaborate with the FDLR to mine gold, cassiterite and coltan deposits. In return for allowing the FDLR access to resources and to tax trade, local communities receive protection and violence is minimized, creating the appearance of peace. However, if local authorities refuse access, the FDLR attacks their communities. This is clearly more a case of extortion than genuine peacebuilding.

Coltan as a conflict mineral

Coltan mining is linked to conflict in the DRC in four ways, each of which is used by armed groups as a strategy to obtain profits from coltan production and trade. The profit strategies are theft, control of production at mine sites, taxation/extortion, and trade. Armed groups involved in coltan production and trade usually engage in more than one of these strategies, and sometimes it can be difficult to distinguish one strategy from another. For example, control of a mine (enabling the armed group to directly sell coltan to *négociants*) usually occurs in conjunction with taxes levied on or extorted from *négociants* transporting coltan they have bought from the armed group away from the mine site.

First, armed groups stole coltan stocks. In the initial months of the Congo War, the Rwandan army and RCD-Goma forces stole between 1,000 and 1,500 tonnes of coltan from the warehouses of the SOMINKI corporation. Congolese miners also reported being attacked and robbed by armed groups. The wave of theft can be explained by armed groups' uncertainty about their future status and tenure during that initial period of conflict. By contrast, as the Rwandan and Ugandan armies and RCD-Goma became ensconced in eastern DRC, their economic incentive was to create conditions that facilitated mining and production that they could then tax into the future, rather than hindering production through theft of equipment or stock. Such strategies are less attractive to individual Mai Mai militias, which have largely been unable to consolidate power in any one area and have faced constant internal divisions and external challenges. For them the best strategy for enrichment has been theft. While some Mai Mai militias do control mines, including coltan mines, it is likely these groups have maintained continuous military control over the area for some period or they have formed a local alliance with another armed group.

Second, armed groups, including both the DRC army and anti-government forces, directly control production. As discussed, some combatants also work as miners. Armed groups controlling the mines may also raise money through selling licences or charging fees to other people to mine the site. In the early 2000s, individual coltan mines in North Kivu and South Kivu were separately controlled by Mai Mai, RCD-Goma, the Rwandan army, the Ugandan army and Rwandan Hutu rebel groups, including the FDLR. The Rwandan army was particularly active in coltan mining, and also had more coltan mines within the territory it controlled, compared to the Ugandan army. Following the peace agreement concluded in 2003, the DRC army returned to eastern DRC and took control of some coltan mines. In 2008, the DRC army jointly controlled mines with both the FDLR and PARECO.

Third, armed groups tax the movement of coltan into and out of territory they control, often through roadblocks or through formal and semi-formal systems of taxation. Taxes were levied on coltan miners and/or traders by Rwandan Hutu rebels (on coltan miners around Walikale in North Kivu during the Congo War), by the FDLR (on the movement of coltan out of the Kahuzi Biéga National Park) and by the CNDP (on goods that passed through the two border crossings that it controlled, the Bunagana and Ishasa crossings). The most comprehensive attempt to extract taxes from the coltan trade was by RCD-Goma. During the Congo War it established a separate civilian administration to run territory under its control, an important step to support its claim of being a state-like entity and to differentiate itself from armed groups motivated purely by economic interests. Needing money to run its administration and to fund its army, and conscious of the very high prices being received by coltan, in November 2000 RCD-Goma imposed a monopoly on the export of coltan. The monopoly was organized by the newly created Société Minière des Grands Lacs (SOMIGL), which was designed to function like a state-owned corporation. SOMIGL was jointly owned by RCD-Goma and three private companies registered in the DRC, one owned by a Belgian, one by a Rwandan and one by a South African. RCD-Goma's objective was to control the entire minerals trade from its territory in order to maximize revenue extraction. The RCD-Goma leadership thought that by monopolizing exports this would ensure all coltan was funnelled through its hands. A tax of $10 was levied on every kilogram of coltan exported, and in the three months following the creation of SOMIGL coltan generated an estimated $2.35 million in taxes.

SOMIGL soon failed, for the same reason other similar schemes in the DRC failed, including the short-lived gold and diamond buying monopoly (or monopsony) established by the government of Laurent-Désiré Kabila in 1998, which was also designed to extract maximum revenue from mineral production. *Négociants* refused to accept SOMIGL's lower prices and withheld their coltan, turning to the black market instead. A couple of *comptoirs* in Bukavu closed their doors altogether. Other comptoirs continued to buy coltan, but stopped exporting. The poor infrastructure and vast size of the DRC made it easy for *négociants* to hide their ore, and porous borders also made it easy for *comptoirs* to smuggle coltan directly into neighbouring countries, bypassing the authorities altogether. When the price of coltan fell in early 2001, it became a less important source of taxes and such low quantities of coltan were being officially exported that SOMIGL's monopsony was abandoned altogether and the market was re-liberalized in March 2001. Still needing revenue, the RCD imposed another form of taxation based on licences, figuring that if it could not properly tax commerce it could at least earn money by requiring annual licences. *Creuseurs* were charged $10, *négociants* $1,000 and *comptoirs* who exported coltan $15,000. Only Congolese citizens were permitted to buy *creuseur* and *négociant* licences, but *comptoirs* could be of any nationality.

The taxation system imposed by armed groups on minerals introduced certainty around investment returns that was sometimes welcomed by both miners and *négociants*. The stage at which porters use *makakos* to carry coltan from mines through jungle and farmland to *négociants* is particularly vulnerable to armed groups, but there appears to be less theft than might be expected. Theft presents a logistical problem in that anyone who stole a heavy *makako* would have to transport it somewhere to be able to sell it. It makes more sense for the armed group simply to extract a tax and allow the porter to continue. Porters also receive protection from the armed groups that have already extracted taxes from them. One *négociant* commented: 'To be honest, it is better for us that [militiamen] are there [in mines]. I can send my buyers walking through the jungle with lots of money, but nobody will touch them as long as we pay the tax. It protects us.'[5] Paying taxes bought protection from predatory behaviour by other armed groups and other miners who might be inclined to rob *négociants*.

Fourth, armed groups became directly involved in the movement of coltan onto the international market, either directly or through part-ownership of *comptoirs*. It was the entry of these 'traders' that alerted established players in the tantalum sector to changes in the structure and control of the industry in Central Africa, even before the United Nations published its reports and NGOs started campaigning. From mid- to late 1999 'several new "traders" of dubious origin were offering parcels of material [ore or concentrates], but in general the established processors were very wary of these newcomers, preferring to stick with trusted long-term suppliers or known producers'.[6] These new traders were 'dubious' because they lacked any commercial provenance in the coltan industry or even in minerals trading. For example, they included a Ugandan airline, a Rwandan electronics company, a Kenyan agricultural commodities dealer and an Israeli gem cutter.

The Rwandan army was prominently involved in this export trade, a role that was facilitated by its ability to illegally transport ore out of the DRC using military networks. The United Nations estimated that in the early 2000s the Rwandan army controlled 60–70 per cent of all the coltan exported from the DRC. Following its official withdrawal from the DRC it became more difficult for the Rwandan army to directly export coltan. However, the CNDP, which received support from the

Rwandan army, continued to work with *comptoirs* in Goma to export coltan in the mid- to late 2000s, and this allowed the Rwandan army to maintain some involvement in the export trade. Following the CNDP's demise and subsequent merger with the DRC army in 2009, some former CNDP military personnel who are now part of the DRC army continue to maintain private business links with Rwandan officials.

Corporate involvement in the coltan trade
The United Nations identified scores of private trading, brokerage, banking and transportation firms as having participated in the illegal exploitation of natural resources from the DRC. Among these firms were twenty-six international minerals trading companies that imported coltan from the DRC via Rwanda during the Congo War. From Rwanda these companies shipped coltan to Malaysia, Germany, Switzerland, the Netherlands, Britain, Kenya, India and Pakistan, from where it was either processed or re-exported to a processing plant elsewhere. Some of these firms were owned and managed by senior figures in the RCD-Goma administration and the Rwandan government.

The aviation companies that transported coltan from the DRC during the Congo War included Alliance Express (49 per cent owned by South African Airways and 51 per cent by the Rwandan government), Kencargo International (20 per cent owned by Martinair), Airflo, Astral Aviation and Martinair Holland (a Dutch firm). Major European firms were also implicated in the illegal coltan trade, including the now defunct Swissair and Sabena and several shipping companies. In the case of Swissair and Sabena, which transported coltan loaded onto their aircraft in Rwanda, there was little doubt most, if not all, of the ore was from eastern DRC. Transportation companies typically claimed that they did not know the origin of the tantalite they were carrying.

The United Nations recommended that financial sanctions be placed on twenty-nine firms, including ten companies from the DRC, Uganda and Rwanda specifically involved in the coltan trade. Another sixteen companies (from Britain, the United States, Germany, Belgium, Switzerland, Malaysia, St Kitts & Nevis and Hong Kong) were also found to be specifically involved in the coltan trade, but were not recommended for sanctions.

The four ways that armed groups have profited from coltan demonstrate that the mineral has contributed to their ongoing viability as organizations. But conflict has also facilitated mining for coltan and other minerals; the cause-and-effect relationship goes both ways. For example, conflict further weakened state institutions, allowing coltan mining to occur illegally and under unregulated conditions in areas where it is not permitted (such as national parks) or not wanted (such as on agricultural land). Conflict also caused people to turn to mining, which was one of the few remaining sources of employment due to the decline of agriculture and trade, historically the mainstays of the eastern DRC economy. Coltan mining was particularly attractive as an alternative to agriculture during the price boom.

Illegal economic networks
The illegal export of natural resources was a major feature of the economy of eastern DRC during the period of Rwandan and Ugandan occupation and was the catalyst for investigations by the United Nations. In its October 2002 report (no. 1146), the United Nations argued that the exploitation was largely carried out by three separate elite networks controlled by senior government and military officials from Rwanda, Uganda and Zimbabwe. What made the trade in Congolese resources illegal is that it is against international law for armed forces occupying a territory to seize resources and

export them. The entire natural resource economy during the Congo War, which depended on the export of commodities out of the DRC onto the international market via the elite networks, was therefore categorized by the United Nations – and the DRC government in regard to exports from occupied eastern DRC – as illegal. The aspect of the illegal export of natural resources that gained the most attention and sparked campaigns by NGOs to regulate the coltan trade was the role of foreign companies. The United Nations' October 2002 report details the relationships between these firms and the elite networks exploiting the DRC's natural resources, and refers to documents obtained by the United Nations that confirm these firms knew their mineral ores originated in the Congo and had passed through the hands of foreign armies.

The United Nations' basic charge was that firms collaborating with elite networks were in breach of the Organisation for Economic Co-operation and Development's (OECD) *Guidelines for Multinational Enterprises*, which provide principles for responsible business conduct (see chapter 4 for more discussion of the guidelines and their principles). Eighty-five firms from twenty countries (including both members and non-members of the OECD) were considered to be in violation of these guidelines because of their collaboration with these elite networks. Among them were several major Western corporations, including Anglo American, Barclays Bank, Bayer AG, Cabot, De Beers, H.C. Starck and Standard Chartered Bank. However, only two of these firms, Cabot and H.C. Starck, were specifically involved with coltan (through their processing plants). Smaller processing plants in Austria, Estonia and Russia may also have received coltan.

One researcher estimates that during the period of the coltan price boom when the Congo War was also at its peak in terms of international involvement, 80 per cent of all coltan was processed in H.C. Starck's German plant.[7] The precise amount of coltan processed by H.C. Starck is unclear, as are the reasons why such a large proportion of coltan may have ended up at its German plant. One possible reason is that the minerals trading firms that historically bought tantalite from Central Africa did not consider the entry of armed groups into the supply chain to be a sufficient basis to change established trading patterns. A second possible reason is that the minerals trading firms dominating the market in Central Africa during the 1990s and early 2000s had links to Europe, so coltan was more likely to end up at a European processing plant, of which H.C. Starck's German plant is the biggest. (In contrast, chapter 5 argues that minerals trading firms from China are now beginning to dominate the tantalite sector in Central Africa.) A third possible reason is that H.C. Starck was slower to ask the difficult questions about whether the ore it was receiving had been handled by armed groups, because senior management were slow to realize the reputational damage that could be caused by continuing to handle tantalite sourced from the Congo during a war. As is discussed in chapters 4 and 5, H.C. Starck's involvement with coltan today has sharply diminished, with the firm announcing that it no longer buys ore it suspects originates from the Congo.

Profits from coltan
During the Congo War, coltan mines were variously controlled by Rwandan Hutu rebels, Mai Mai, the Rwandan army, the Ugandan army and RCD-Goma forces. The majority of coltan mines that were controlled by armed groups were controlled by the Rwandan army and RCD forces, and most coltan profits went to these organizations. As armed groups have emerged and merged, however, so too has their control over coltan mines and trade evolved. In 2008, the CNDP, FDLR, DRC army and Mai Mai forces (i.e. both pro- and anti-government forces) profited from coltan. Following the defeat of the CNDP and its absorption into the DRC army, ownership has changed again.

In 2009, the best available estimates[8] were that the FDLR fully controlled up to six coltan mines (all in South Kivu); one coltan mine was jointly controlled by FDLR and Mai Mai forces (Ibanga mine, North Kivu); two coltan mines were fully controlled by the DRC army; and ten coltan mines were controlled by former CNDP forces that now came under the umbrella of the DRC army, although the extent to which non-former CNDP troops could actually enter the mines or tax miners is unclear. Another three coltan mines in North Kivu and one mine in South Kivu were reportedly not controlled by an armed group.

There is no doubt that armed groups have used some of the profits from coltan mining to fund their activities. Adolphe Onusumba, a former leader of RCD-Goma, famously claimed in 2001: 'We raise more or less $200,000 per month from diamonds . . . coltan gives us more: a million dollars a month.'[9] However, estimating armed groups' fiscal reliance on mineral resources is a difficult and inexact task, as is distinguishing profits from coltan from the profits from other minerals. In addition, not all armed groups rely on natural resources to the same degree. For example, by one estimate the CNDP derived no more than about 15 per cent of its revenue from the mineral trade, while the FDLR derives up to 75 per cent (mostly from gold) and the DRC army's 85th brigade up to 95 per cent.[10]

Data on profits extracted from coltan by armed groups is not kept systematically and is not available for most armed groups. However, research has produced some estimates of coltan-related profits for all armed groups operating in the DRC and these are shown in Table 3.3. The high figure for 1999, $62.6 million, is undoubtedly due to the looting of SOMINKI's coltan stockpiles shortly after the commencement of the Congo War and before the price boom.

As there is presumably some correlation between the value of exports of a particular mineral and the profits it generates for an armed group, comparative data on the value of

Table 3.3	Estimated profits from coltan mining, selected armed groups[11]		
	1999	*2000*	*2008*
Rwandan army	$62,600,000		
Rwandan army and RCD-Goma (combined)		$10,000,000	
All armed groups			$11,800,000

Source: table compiled by author using data for 1999 from United Nations (2001), *Report of the panel of experts on the illegal exploitation of natural resources and other forms of wealth of the Democratic Republic of the Congo* (Report S/2001/357); data for 2000 from Martineau, Patrice (2003), *La route commerciale du coltan: une enquête.* Groupe de recherche sur les activités minières en Afrique, Faculté de science politique et de droit, Université de Québec, Montréal; and data for 2008 from Enough Project, with the Grassroots Reconciliation Group, Washington DC (2009), *Comprehensive approach to Congo's conflict minerals.*

different minerals can be used to indicate coltan's importance as a source of income relative to other minerals. Data on the value of gold, tin and coltan exports from South Kivu are available for 2005; and data on the value of gold, tin, coltan and tungsten exports from South Kivu are available for 2008 (see figure 3.2). As the data are only for South Kivu it does not give information about profits made by armed groups from coltan exported from other provinces, notably North Kivu.

The data in figure 3.2 can be assumed to underestimate coltan profits as it is taken from the value of official exports recorded by the DRC's customs agency, the Office des Douanes et Accises, and due to corruption and smuggling many exported minerals are never officially registered. Nevertheless, the numbers suggest that profits from coltan exports are much less important as a source of revenue compared to gold (nine times more valuable than coltan in 2005, and four times more valuable in 2008) or tin (seven times more valuable than coltan in 2005, and 2.5 times more valuable in 2008). However, coltan exports were 3.3 times more valuable than tungsten in 2008.

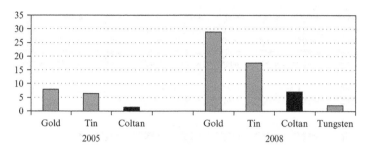

Source: Data for 2005 are from the DRC's customs agency, the Office des Douanes et Accises, as published in Hayes, Karen, Hickock Smith, Kimberly, Richards, Simon and Culp Robinson, Richard (2007), *Researching Natural Resources and Trade Flows in the Great Lakes Region*. PACT, Washington DC, p. 21. Data for 2008 are from the Enough Project, with the Grassroots Reconciliation Group (2007), *Comprehensive Approach to Congo's Conflict Minerals*. Washington DC.

Figure 3.2 Estimated value of mineral exports from South Kivu, millions of dollars.[12]

Notwithstanding uncertainty regarding coltan profit data, it is interesting to calculate the contribution coltan profits may make to fuelling violence in eastern DRC. A useful measure is the number of weapons that could have been purchased with coltan profits, and this can best be done using African armed groups' weapon of choice: the AK-47. The AK-47 is a Soviet-designed machine gun renowned for its robustness, reliability, ease of use and cheap price. Figure 3.3 shows the number of AK-47s that could have been bought by armed groups using profits from coltan.

Social impact of militarized coltan production

Armed groups involved in coltan production and trade in the DRC are directly implicated in atrocities against civilians, including sexual violence and extrajudicial killings. Armed

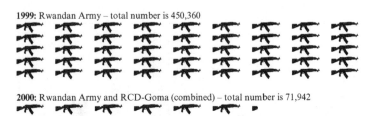

1999: Rwandan Army – total number is 450,360

2000: Rwandan Army and RCD-Goma (combined) – total number is 71,942

2008: All armed groups – total number is 84,514

Source: Table compiled by author using profit data for armed groups shown in Table 3.3. Price data for AK-47s are from Killicoat, Phillip (2006), *Weaponomics: The economics of small arms.* Working Paper, Centre for the Study of African Economies, Oxford University, Oxford, table 2, p. 7.

Figure 3.3 AK-47s that could be purchased, from coltan profits; selected years. One gun represents 10,000 AK-47s.

groups have brutally recruited civilians to assist them, including forcibly recruiting child soldiers. It is these atrocities that have galvanized international efforts to target the economic activities that have funded armed groups.

Chapter 2 argued that artisanal and small-scale mining of coltan (and other minerals) attracts armed groups because of the opportunities it creates, and that armed groups can get away with using violence in artisanal mines because labour is cheap and lacks collective bargaining power. These arguments highlight the relationship between violence and certain kinds of mining methods, and therefore the relationship between violence and certain minerals, given that some minerals lend themselves to artisanal mining while others do not. However, it is impossible to point to coltan-related profits and argue that these profits have enabled a particular armed group to commit specific abuses, or that armed groups that profit from the coltan industry are more violent than armed groups that profit from other non-mineral revenue-raising activities, such as agriculture, logging or looting. To be able to demonstrate that there is a direct causal effect between the coltan industry and violence by armed groups it would be necessary to show that:

- violence happens more frequently or with greater intensity near coltan mines than it does in non-mining areas;
- violence happens more frequently or with greater intensity along primary coltan trade routes than non-mineral trade routes;
- armed groups engaged in the coltan trade commit proportionally more acts of violence than those not engaged in the coltan trade; and
- armed groups that control more coltan mines or make more money from coltan engage in proportionally higher levels of violence than those who control fewer mines and make less money.[13]

To date, there is no evidence to demonstrate that the conditions created by coltan mining are any better or worse than conditions created by other economic activities in which armed groups engage in eastern DRC, including both the exploitation of other minerals and non-mining activities. Nevertheless, there is plenty of evidence that all armed groups engage in violence as part of their revenue-raising activities.

Sexual violence

All armed groups, including the DRC army, are implicated in sexual violence in eastern DRC. Sexual violence includes rape, sexual slavery, mutilation of genitalia (including women being shot or bayoneted in the vagina and men being castrated), and forcing family members to participate in rape or mutilation of genitalia. Some victims of rape are subsequently murdered. A 23-year-old woman from South Kivu, Mapendu, described the time she was raped:

> It was during the night. About thirty of us were in the church doing our night prayers. Then about thirty-seven rebel troops

> broke in [and took us to the forest] . . . It felt like it went on forever. I was in so much pain. It took me four hours to walk home. It normally takes me just 45 minutes.[14]

A couple of years later Mapendu was raped a second time, this time by soldiers from the DRC army.

The consequences of this violence include psychological trauma (e.g. depression and suicide), physical trauma (e.g. fistula, broken bones, severed limbs, burns, gunshot and stab wounds), medical effects (e.g. sexually transmitted infections, HIV transmission, tetanus, hepatitis B and unwanted pregnancy), social effects (e.g. family breakdown, rape survivors being rejected by their families) and economic effects (e.g. women and girls being unable to work). It is mostly women who are raped, but rape of men appears to be an increasing phenomenon and one clinic in Goma reported that, in June 2009, 10 per cent of rape survivors were men.

Data on rape are not comprehensive because of the difficulty obtaining them. However, its magnitude is unequivocal. In just one province (South Kivu) in one year (2006) the United Nations estimated that there were 27,000 rapes. In 2007, a United Nations survey of half the health centres across the DRC found that 50,000 rapes had been reported; for the month of June in 2008, 2,000 rapes were reported in North Kivu alone (assuming this is an average, that would mean 24,000 rapes over the year); and a humanitarian NGO, the International Rescue Committee, reports that it has assisted over 40,000 rape survivors in the DRC. In 2009, a doctor for Médecins Sans Frontières reported receiving at her clinic about 20 women who had been raped per week – previously it had been 50 a week. The United Nations estimates that at least 200,000 women and girls have been the victims of sexual violence in the DRC since 1998 although, given that rape is not always reported, this figure is surely an underestimate.

With the dissolution of some armed groups over the course of the 2000s sexual violence is now committed by fewer groups, but the rate of violence appears to be constant. In 2009 the forces of the FDLR and the DRC army were responsible for most of the sexual violence that occurred, although one report estimates that the DRC army is the no. 1 perpetrator.

Extrajudicial killings
Extrajudicial killings – that is, organized execution that is unrelated to any legal civilian process as the result of an alleged breach of law – have resulted in the deaths of tens of thousands of civilians. All armed groups operating in eastern DRC have carried out extrajudicial killings. Rwandan and Burundian refugees were the main targets of extrajudicial killings in 1996–7, but during and subsequent to the Congo War Congolese civilians – including children and babies – have been the main targets, often on the basis of ethnicity. As with other forms of violence, North Kivu and South Kivu are the worst-affected provinces.

Child soldiers
All armed groups have recruited child soldiers. An estimated 30,000 children have fought as combatants in the DRC since 1998, and as recently as 2009 there were up to 8,000 child soldiers in the ranks of armed groups. The recruitment of children is not benign. Armed groups kidnap children from their families, force them to participate in training and indoctrination sessions, and then prevent them from leaving camps, threatening them with death if they escape. Child soldiers have also been forced to participate in sexual violence and killings of civilians.

One former child soldier from South Kivu, Bintu, reflected on her experiences in the DRC army, which she voluntarily joined when she was 17:

I killed five people. When I think about it I feel really bad, but nothing would deter me because I was just doing my job. Before going out on an operation we would be high on drugs . . . If they [civilians] didn't give us what we wanted it would make them [the other soldiers] angry. They would rape the women and I would help them as they did it . . . I would definitely go back to that life. I haven't got anyone in Bukavu. I don't have any means of survival. Nowhere to live, nowhere to sleep. I have to prostitute myself to survive.[15]

At the time Bintu told this story she had a 4-year-old son, whom she hoped to place in an orphanage so she could return to the army.

Social conditions around coltan mines
It is clear from atrocities committed by armed groups against civilians that coltan mines and their hinterlands can be desperate places. Lots of money is made by a small number of people, but most miners work long hours for little pay in conditions of uncertainty and danger created by landslides, collapsing pits and violence at the hands of armed groups. Inhabitants of coltan-producing areas in 2000 reported extensive prostitution, alcohol and drug use and breakdown of social mores. HIV/AIDS is probably widespread, though likely to be undiagnosed.

As occurs elsewhere in the world, a close relationship between the sex industry and mining has developed. In one mine, sex could be bought for a dessertspoon of coltan (worth about $3.75 at the time). In another mine of 300 miners and 37 prostitutes, miners paid a kilogram of coltan (about $80) to a madam to gain the right to indefinitely 'pair off' with a prostitute. The miner could end the arrangement when he decided he wanted a new prostitute, as could the prostitute if she decided she wanted a new miner (better at producing coltan). In both cases a fee for infringing the arrangement was payable to the madam – another kilogram of coltan. The madam's financial incentive was to encourage as much bed-hopping as possible. Sexually transmitted diseases were rife.

Inhabitants involved in coltan production and trade in North Kivu were interviewed by the Goma-based Pole Institute in 2000. They made the following comments about coltan and life in coltan mines during the Congo War:

> It makes miners proud and arrogant, like large cattle owners previously.
>
> Faustin Ntibategera, director of local NGO

> We think that agricultural activities are a good thing, but we cannot see ourselves taking them up again in the short term because we earn much more money from coltan. However, we are thinking of investing coltan money in agriculture and cattle once peace returns.
>
> anonymous miner

> Coltan mining is very profitable, but only the husbands profit from it. Once they have the money, they go away and look for other women in Goma for whom they even buy houses while our own children suffer and don't go to school.
>
> anonymous housewife

> Coltan has at least solved the unemployment problem. That has significantly reduced theft. Also many young fighters have turned into coltan miners. That also reduces their number and that of murder. Having said this, there is more money and so the price of foodstuffs has risen.
>
> Safari Lupfutso, chief of Matanda, Buhunde commune

> There is moral depravity in the mines: no morals, no difference between the sexes in this activity where you find men and women working naked. Prostitution is booming in these immoral mines, couples are formed just like that ... There is drug and alcohol abuse even among the children.
>
> Emmanuel Mulindwa, priest, Matanda village

> Landslides cause many victims. Often we have to dig up to 6 metres deep on hillsides and normally up to 3 metres in the

valleys. Sometimes soldiers take our produce on the road. Often our employers cheat us on the sales price and give us hardly anything.

Nzakura and Mdagije, miners in Luwowo and Mishavu

Coltan is an unprecedented source of income for many men and women, so that entire households come to live in the mines and keep their children out of school.

Christine Kizimana, leader of a local NGO

In general the armed groups do not bother the general population as long as the latter keep away from the mining.

Bashali Bokalos, Baptist Church, Mashali Mokoto village, Masisi

These resources attract all warring parties who fight for their control. The indigenous population which has access to the mines concentrates mainly on drink and prostitution instead of investing the profits they make like they should.

Nyanguba Mwicho, chief of Ihana, Wanyanga commune, Walikale

There are contradictions in these comments. Some people argue that coltan contributes to violence; others argue that it distracts belligerents from violence; men seem to profit more from coltan than women. The divergent views can probably be accounted for by the fact that interview data were collected in different sites from a range of people with very different roles and positions in the mining community. While coltan mining has had some general effects on surrounding communities, such as contributing to the demise of agriculture and attracting people with limited economic opportunities, there has been variation in these effects over time and location.

The broad themes in descriptions of social conditions in 2000 continue today. In 2010, mining offers some of the few economic opportunities in the region, and riches for those who strike it lucky. However, accidents and deaths remain common, prostitution and drug and alcohol consumption are widespread, mines still attract children who run away from school and parents, men abandon families to work in the mines, and mining communities lack the social fabric and social capital of Congolese communities outside the region.

Mining for coltan and other minerals has had some specific impacts on forest dwellers. Bambuti (Pygmy) communities living in and around the Okapi Wildlife Reserve in Orientale province harvest honey from eko trees, the flowers of which attract bees that make hives and honey in the trees. Miners' preference for the bark of eko trees to make troughs or sluices in which coltan is separated out using water has caused thousands of these trees to die from being ring-barked, depriving Bambuti communities of a food source.

Communities outside eastern DRC have not suffered from mining in the same way. Even during the height of the Congo War in 1999 and 2000, conditions away from the front line remained relatively stable. Most Congolese had very low incomes but their lives followed rhythms associated with work, school, religion and family that are recognizable to people around the world.

Conclusion

Violence in the DRC is not the result of a single issue, and the coltan industry is no more culpable than any other natural resource in the perpetuation of conflict. It may be the case that for some period the RCD-Goma earned 'more than a million dollars a month' as the RCD-Goma leader, Adolphe Onusumba, boasted. The Rwandan army also profited handsomely from the coltan it stole and the mines it occupied during the Congo War. However, this period of great profitability was confined to the coltan price boom. The fact that coltan is no longer so important complicates calls to regulate coltan production and trade in order to reduce armed groups' ability and motivation to wage war. Armed groups that do not profit from coltan mining would be little affected by regulation of the industry unless the new rules and procedures were extended to all minerals, and may well continue fighting for the reasons they decided to take up arms in the first place. The complex and evolving causes of violence are, of course, the reason conflict in the DRC has been so intractable.

This chapter analysed the role of coltan and other minerals in continuing violence in the DRC, but the importance of coltan as a causal factor in the war is not actually what has made the mineral politically significant. In twenty years' time, the price spike in 2000 and the subsequent coltan rush by civilians and armed groups may well be viewed as a minor chapter in a complicated war. The reason coltan has become significant is the way it has been used by activists to draw attention to a war neglected by the international community. Coltan has become symbolic of how ordinary people on the other side of the world, through their consumption habits, are implicated in conflict and injustice. In a trend exemplified by the campaign against blood diamonds, advocacy groups have developed initiatives that explicitly target armed groups' exploitation of natural resources, by drawing attention to the consumption of end products containing those resources. Activists engaged in coltan advocacy have refined the tactics and marketing used in such campaigns, including linking the exploitation of natural resources to sexual abuse, environmental destruction and threats to endangered wildlife. The United Nations first documented the role of coltan in the Congo War, but it was activists who identified opportunities created by those reports to manipulate governments and to put pressure on corporations wishing to ignore the possibility they might be implicated in the conflict.

4

Advocacy, campaigns and initiatives

The politics of natural resources were once determined by corporations seeking profits and governments seeking taxes. Affected communities had no say in the timing, duration or pattern of exploitation; indigenous communities had no hope of influencing anything. The situation today could not be more different, with civil society NGOs, multilateral organizations and communities acquiring power and influence in the natural resources sector. At the forefront of coltan politics are NGOs determined to reshape the coltan trade as a way of influencing war in the DRC. Many of the organizations involved in campaigning about coltan were already engaged with the Congo, including on conflict, sexual violence, human rights and development issues.

This chapter analyses ten advocacy campaigns and initiatives focused on coltan. It also analyses seven other initiatives designed to have a general impact on the conditions under which natural resources are exploited. The ten initiatives with a specific focus on coltan share a common broad objective: raising awareness of the impact of continuing conflict on the DRC. Most of these initiatives also aim to stop armed groups in the DRC profiting from coltan mining and trade. Two of them seek to halt coltan mining in a national park to help protect a species of gorilla.

The challenges facing these coltan campaigns are analysed, including the difficulty of marketing coltan as a social cause. Coltan does not become beautiful jewellery, unlike diamonds and gold, and cognisant of coltan's lack of charisma NGOs have made mobile phones – a globally ubiquitous accoutrement that people depend on and care about – the focus of their initiatives. Another problem for NGOs seeking support for their coltan activities is competition from other issues. The global justice 'marketplace' is crowded and NGOs advocating on coltan and the Congo face competition from blood diamonds, 'dirty gold', Darfur, Burma, Tibet, Palestine, the Amazon, fair trade, Third World debt, imprisoned writers, slavery, gorillas, elephants, whales and climate change. Some coltan initiatives link coltan and the Congo to these other more established issues, such as gorilla survival and destruction of the rainforest.

Campaigns and initiatives

There is ongoing debate in the scholarly, activist and international community about the best way to stop armed groups profiting from the coltan industry and other conflict minerals. Four basic interlinked issues have been identified: involvement by armed groups in the production and trade of minerals; the weakness of the DRC state, which is unable to control anti-government armed forces; corporations willing to buy minerals that have been handled by armed groups; and the need for alternative livelihoods for civilians, other than mining. This section analyses the ten campaigns and initiatives that seek to change aspects of the coltan industry, including their objectives, rhetoric, marketing and implementation.

The initiatives are grouped according to the stage in the supply chain on which they primarily focus: production, trade and/or demand (including for tantalum base products and retail products containing tantalum). Table 4.1 lists the initiatives and the perceived problem area associated with

Table 4.1 Coltan initiatives and their supply chain focus

	Production		Trade		Demand	
	Mining practices	Property rights: land tenure	Domestic provenance	International provenance	Corporate buying practices	Consumer preferences
Durban Process for Ethical Mining	✓	✓				
United Nations initiatives			✓	✓	✓	
Certification system			✓	✓		
Fingerprinting coltan ores			✓	✓		
Enough Project				✓	✓	✓
US legislation			✓	✓	✓	
No blood on my mobile					✓	✓
Break the connection					✓	✓
Breaking the silence						✓
They're calling on you						✓

Source: compiled by author using information from initiatives.

production, trade and demand that each aims to address. What is striking is the lack of initiatives focused on the upstream (production) end of the supply chain which, because it involves working in the DRC with miners, *négociants* and the government, is also the more difficult end. NGOs concentrate their efforts on trying to change corporate and consumer demand because this is where NGOs have the most leverage and experience. Boycotts of mobile phones and not sending text messages for a certain period, as a means of highlighting and protesting the presence of 'conflict tantalum' in mobile phones, have become particularly popular. For example, in 2009 the Catholic Church in Italy launched a 'No SMS for Lent' initiative, and asked Italians to pause for reflection during Lent as a 'sign to remember that, as that [text] message is written, we are shaping the lives of people far away'.[1]

Initiatives focused on production

Durban Process for Ethical Mining
The Durban Process for Ethical Mining ('Durban Process') is the sole initiative to focus on mining practices and security of property rights, in this case in a national park. Its goal was to preserve and protect the habitat of the eastern lowland gorilla, or Grauer's gorilla, 85 per cent of which live in Kahuzi Biéga National Park. The eastern lowland gorilla is critically endangered as a result of habitat loss and poaching in the park. The main threat comes from illegal mining – operations destroy forest cover and miners kill and eat gorillas and other wildlife. Most mining within the park is for cassiterite, but tungsten, cobalt and gold are also mined. Coltan is mined on the park's northern edge. Implemented between 2003 and 2008, the Durban Process was the initiative of European environmental NGOs, spearheaded by the Great Britain-based Gorilla Organization. It was funded by the United Nations,

the International Union for the Conservation of Nature, the Communities and Small-scale Mining Secretariat of the World Bank, and private charities in Europe and the United States. H.C. Starck was a member and contributed funding. Local Congolese, including local government authorities, were involved in the design and management of the project. The project was concluded in 2008 when it was unable to obtain funding to continue its activities. The project aimed to disseminate the DRC government's mining code among miners; raise awareness on environmental issues through training as a way of making mining practices more ethical; develop alternative livelihoods to mining; and support the Institut Congolais pour la Conservation de la Nature (such as through training parks rangers and equipping patrol posts). The project also provided training in alternative livelihoods (livestock rearing and farming) for miners' wives.

Training in ethics and codes of conduct are common initiatives that organizations use in an attempt to obtain compliance with policies and procedures. However, to be successful incentives are required to get recipients of training to comply with the standards and to make ethical decisions, including sanctions if they do not comply and rewards if they do. Examples of incentives might include higher income for ethical miners, but punishment if they breach the code of conduct. For the project to be successful, coltan produced ethically outside the national park would probably have to receive higher prices than coltan mined inside the park, and this might involve some collaboration with *négociants*. Obtaining compliance with a code of conduct also requires oversight. In most workplaces this comes in the form of managers, whose job it is to make sure staff behave the way they should. Mining communities have such figures: the *chefs de groupes, chefs de chantiers* and *chefs de collines*. These 'managerial' figures also need incentives and pressure from above to encourage them to implement procedures. Like other communities in the DRC, poverty in artisanal mining communities is widespread. Miners would need to be over an income threshold (regardless of source) and have their basic human needs met, before they would entertain the idea of adjusting their practices in a way that might result in loss of income. Finally, there is the problem of the commons: if ethical miners desist from mining in certain deposits, other miners may occupy those areas. The project came up against this problem early on when miners who left the park, following their training, returned when they discovered their colleagues who had remained went unpunished for mining illegally. The Durban Process subsequently recognized that collaboration with government authorities (e.g. police) would have been necessary to take action against non-compliant miners.

Initiatives focused on domestic trade

United Nations reports and initiatives
Much maligned for the apathy of its member states about Congolese affairs, the United Nations has nevertheless played an important role in putting coltan onto the public's radar screen. United Nations initiatives include two measures that have a direct impact on the coltan supply chain. MONUC has commenced random checks of aircraft and boats to detect illegally produced and traded resources. It has even made some seizures, although these were tiny. Such interdiction increases the threat to traders and anti-government armed groups handling ore with a suspicious provenance. The United Nations has also adopted a regime of sanctions, which are discussed below in conjunction with the resolution that introduced the sanctions.

The major contribution of the United Nations to coltan politics has been in raising the profile of violence and the illegal exploitation of natural resources in the DRC, and this section focuses on the reports and resolutions that have achieved this. A series of reports by the United Nations Panel of Experts on the Illegal Exploitation of Natural Resources and Other Forms of Wealth of the Democratic Republic of Congo in the early 2000s, and then the Group of Experts on the Democratic Republic of Congo in the late 2000s, 'named and shamed' corporations, armed groups and individuals. The reports' findings were supported by detailed documentation of trade relationships between Congolese, regional African and other international actors, especially private companies. It was the Panel of Experts' initial report released in April 2001 (no. 357)[2] that triggered the 'no blood on my mobile' campaign by Belgian NGOs angered by the involvement of Belgian companies.

The rationale behind the creation of the Panel of Experts and the emphases of early reports may have been biased by the interests of UN Security Council members. For example, the April 2001 report (no. 357) was strongly critical of Rwanda and Uganda, ascribed to the influence of France, and the November 2001 report (no. 1072) was strongly critical of Zimbabwe and omitted criticism of British and American companies due to the alleged influence of Britain and the United States. However, as a series, the reports document the illegal activities of all parties and make valid findings. In regard to natural resources, the terms of reference for the panels were general in that they covered any and all natural resources, including coltan. The November 2001 report (no. 1072) had a specific section documenting the coltan industry. The January 2007 report (no. 40) is notable because it was the first time the United Nations called for sanctions based on activities deemed illegal under DRC law. The December 2008 (no. 773) and November 2009 (no. 603) reports update previous analyses with information about armed groups active in

2008. They have extensive sections on the CNDP, the FDLR and two other minor armed groups, and these groups' exploitation of natural resources, involvement in illegal trafficking of raw materials and arms, and continuing violence. Coltan is given specific attention.

There have been no UN Security Council resolutions specifically focused on coltan, but following publication of the April 2001 report (no. 357) coltan certainly became one of the natural resources on the Security Council's agenda. Interpreting Security Council resolutions and linking their timing to political events is an exercise in language. The Security Council is restrained in its choice of words, uses a limited vocabulary to express varying stages of concern and the subjects of resolutions are often obliquely described. Dozens of resolutions on the DRC have mentioned natural resources, but five resolutions are most important because of their linking of coltan to conflict: resolutions 1291, 1355, 1457, 1592 and 1856.

Resolution 1291 (in 2000) was the first to mention the exploitation of natural resources in connection with the Congo War. It notes the Security Council's 'concern' at reports of the 'illegal exploitation of the country's [DRC's] assets and the potential consequences of these actions on security conditions and the continuation of hostilities'. Following the release of the April 2001 report (no. 357), the language of the resolutions becomes stronger and the exploiters themselves are targeted. Resolution 1355 (in 2001) notes that the report 'contains disturbing information about the illegal exploitation of Congolese resources by individuals, Governments and armed groups involved in the conflict and the link between the exploitation of natural resources . . . and the continuation of the conflict'. In a reference to Rwandan and Ugandan forces occupying Congolese territory, it also reaffirms the 'sovereignty' of the DRC over its natural resources. Prior to resolution 1376 (2001) the Security Council had 'condemned' the actions of armed groups, but in this resolution – timed to coincide with the Group of Experts' second report – armed groups' *illegal exploitation of natural resources* is also 'condemned'.

With resolution 1457 (in 2003) the relationship between the exploitation of natural resources and violence is brought to the fore. The resolution states that 'plundering' of natural resources is 'one of the main elements fuelling the conflict in the region' and reiterates that they should be 'exploited transparently, legally and on a fair commercial basis, to benefit the country and its people'. Later in 2003, resolution 1493 announces an arms embargo on illegal armed groups in the DRC and gives MONUC a mandate to enforce the embargo. It also announces the Security Council's 'intention to consider means that could be used to end' the illegal exploitation of natural resources. The significance of resolution 1493 for coltan is that the Group of Experts' reports found that armed groups illegally exploiting natural resources were using the profits to buy weapons. For the first time the illegal exploitation of natural resources was directly linked to armed groups' capacity to wage war, as were private firms that transported minerals out, and carried weapons into, the DRC. Resolution 1592 (in 2005) finally directly implicates neighbouring states in the illegal trade of natural resources. It urges states neighbouring the DRC (i.e. Rwanda and Uganda) to 'impede any kind of support to the illegal exploitation of Congolese natural resources, particularly by preventing the flow of such resources through their respective territories'. With resolution 1756 (in 2007) the DRC government's own forces are finally mentioned as well. The Security Council urges the DRC government to improve 'the transparency of the management of the revenues' from natural resources. This is an oblique reference to the DRC army and other officials corruptly profiting from illegal mining and logging. Resolution 1794 (in 2007) more explicitly links Congolese actors to illegal activity. It urges states in the region, 'including the Democratic Republic of the Congo itself, to take appropriate steps to end the illegal trade in natural resources'.

Resolution 1856 (in 2008) expanded MONUC's mandate in two important ways. First, MONUC was permitted to conduct joint operations with the DRC army to prevent the 'provision of support to illegal armed groups, including support derived from illicit economic activities'. Second, MONUC was permitted to 'use its monitoring and inspection capacities to curtail the provision of support to illegal armed groups derived from illicit trade in natural resources'. These were major changes to MONUC's operation, and gave real teeth to previous resolutions about illegal exploitation of natural resources. Simply by trading coltan and other minerals, illegal armed groups could make themselves targets of MONUC's forces.

In a country where the government is incapable of enforcing the rule of law, where policies and regulations change and are applied capriciously by officials, and officials on a daily basis engage in illegal corrupt behaviour by demanding bribes and engaging in extortion, discussing the economy in terms of legal and illegal acts as the United Nations and some analysts do can be an impediment to analysis. The informal, unregulated economy has always been a feature of the DRC and also sustains most Congolese. Informal economic behaviour, including smuggling, is a survival mechanism in the face of avaricious officials. The issue for the United Nations was that illicit economic behaviour during the Congo War involved *states*, such as Angola, Rwanda, Uganda and Zimbabwe, among others, via their militaries which were participants in the conflict. As the United Nations is an organization comprised of member states, and as its members have a primary interest in controlling what travels across their borders, the international dimension of the informal and illegal exploitation of natural resources in the DRC came under its scrutiny.

Under pressure to define precisely what it meant by 'illegal' when it talked about the exploitation of natural resources in the Congo, the United Nations came up with a four-part definition: violations of sovereignty, belligerents' (lack of) respect for existing regulatory frameworks, discrepancies between widely accepted practices in trade and business and what occurs in the DRC, and the violation of international law.

Have the United Nations reports and resolutions achieved anything? The Congo War raged on for three years after the first resolution and two years after the first report, and local-level violence continues today. Resolutions and reports and their findings are worthless without member state support and commitment to action, and lack of support and commitment have been one of the main constraints on international peace and development initiatives in the DRC. The resolutions have also largely ignored the role of the DRC government armed forces in the illegal exploitation of natural resources in that they have not explicitly condemned the DRC army, always referring to the activities of '*illegal* armed forces' to differentiate groups such as the FDLR from official government forces. But like the CNDP or FDLR, the DRC army's 85th brigade also lacks official rights to the coltan deposits it controls. Notwithstanding their limitations, reports and resolutions by the United Nations can put an issue into the public arena, get governments (and NGOs) to focus on an issue, and identify targets for pressure and change. A measure of the importance of the work of the United Nations is that advocacy groups continue to cite its reports and resolutions when exhorting governments to act against companies involved in the coltan trade.

Resolution 1856's call for a 'plan for an effective and transparent control over the exploitation of natural resources including through conducting a mapping exercise of the main sites of illegal exploitation' is also being translated into action by donors. The German government is funding projects to 'fingerprint' coltan ores and develop a certification system for small-scale mining of minerals in eastern DRC. The British government is funding the International Peace Information Service in Belgium to map areas of militarized mining in North Kivu and South Kivu.

Improving the management and governance of natural resources is also an objective of an integrated security, economic and social development plan to stabilize and rebuild eastern DRC that has been agreed to by MONUC and the Congolese government. A key element of the plan is to create central buying and transportation points for natural resources at the provincial level, including at existing airports, ports and border crossings. The idea is to locate state authorities where minerals and timber need to transit, to make it easier to collect official taxes, compile statistics and certify the origins of minerals, as well as making it more difficult for criminal groups to transport commodities. Whether the centres are utilized will depend on whether there are sufficient incentives for *négociants* and *comptoirs* to use them. The plan's reliance on checking and taxing by bureaucrats is a risk to the efficiency, integrity and robustness of the system.

Certification system
Germany's Ministry for Economic Cooperation and Development is funding a project to develop and implement a national certification system for small-scale mining of tin, tantalum, tungsten and gold ores in eastern DRC. The project is part of a wider German government-funded initiative to combat illegal exploitation and trade of natural resources in the African Great Lakes region. The project is scheduled to run from late 2009 to mid-2012. It aims to implement a pilot certification system in a discrete area, probably South Kivu, in 2010 to test-run the system. In addition to developing a

certification system, the project aims to build the capacity of DRC state institutions to collect mine production data and play a regulatory role, as well as link the system to the DRC's mining code to ensure mineral production complies with laws and regulations. The objective is to prevent armed groups profiting from artisanally mined ores by excluding from the supply chain the minerals they produce or trade.

Establishing an effective certification scheme for coltan will face large, if not insurmountable, challenges. Coltan is produced in many different artisanal mines and if the proposed certification scheme is to be successful different consignments of ore will need to be kept separate. Until fingerprinting technology can be used with confidence to distinguish the origin of ores from different sites within the Kibaran deposit, *négociants, comptoirs* and international minerals traders will be tempted to mix illegally produced or traded ore with certified ore, or to substitute consignments of certified with illegal ore. Given the porous border between the DRC and Rwanda, and given that the latter also produces tantalite, it is likely that coltan will also be smuggled into Rwanda and certified as Rwandan tantalite. Cooperation from the DRC government will be essential to the success of the pilot scheme. Yet, any attempt to refuse certification of coltan from mines controlled by the DRC army will probably create consternation among Congolese officials, politicians and military leaders.

Preventing mixing and substitution will require hard controls, including containers that are difficult to breach (the old flour sacks typically used in the Congo to transport bulk materials are not appropriate for the job), isolation of consignments in lockable depots and, most importantly, seals. Seals need to be cheap, low-tech, easy to deploy, difficult to defeat or corrupt, yet easy to verify. Such seals exist – they are widely used for nuclear material, for example – so the difficulty is not obtaining them. The biggest problem is likely to be identifying a trustworthy group of people to whom seals can be distributed, and then maintaining an inventory of where they are located.

A workable certification system for bulk material such as coltan also needs a 'release from certification' point, and it is not yet clear from the information available where this point would be. A release point is needed because a certification scheme that continued to trace coltan once it has been delivered into administratively competent entities, such as manufacturers of electronics components or alloys, would add little or no value. After the complex processing that chemically dissolves tantalite as a precursor to turning into it tantalum base products, it would also be impossible to distinguish the origin of the tantalite, making any certification scheme redundant. This means that certification should probably focus on the upstream area, where coltan is extracted, bagged and traded. The problem with such a focus is that coltan mines are remote and, under current conditions, highly insecure. Nevertheless, a scheme that traces coltan from the production stage and releases it once it has been processed, is likely – when conditions permit – to be the most workable certification method.

Fingerprinting coltan ores

A research project to chemically 'fingerprint' coltan ores is being undertaken by scientists from Germany, Canada and Belgium. It is funded by the German government and supported by the DRC government. The project, commenced in 2006, aims to develop technology that can reliably distinguish the origin of tantalum ore concentrates. Such technology will be an important complementary tool to any certification scheme for tantalum ores. The fingerprinting project has made some progress. Scientists have been able to identify and differentiate the mineralogical and chemical characteristics of tantalite ores from the DRC, Rwanda, Mozambique, Ethiopia and Namibia (statistically significant with 90 per cent confidence).

The fingerprinting process must be replicable, practical and reasonably priced if it is going to be adopted by industry. Importantly, early results indicate that the project's methods and technology can be replicated by laboratories elsewhere. But there are some technical challenges that need to be resolved. The quality and composition of coltan ore may vary depending on the equipment used to extract the ore, and the skill of the miners. This may mean that not all samples can be reliably tested. Certification through fingerprinting must also not significantly raise the price of coltan compared to tantalite from elsewhere.

Furthermore, while being able to identify and differentiate the mineralogical and chemical characteristics of ores with '90 per cent confidence' may sound impressive, this means there is a 10 per cent chance a tantalite ore will be identified as being from the wrong country. The technology would be considered more robust if it obtained results that are statistically significant with 95 per cent confidence – the standard level considered acceptable in statistical analysis. The results are further complicated by regional and local variations in the composition of coltan within the DRC. The current technology does not allow identification of locations down to the district level. Therefore, while Congolese coltan can be distinguished from Rwandan tantalite with some confidence, the technology cannot distinguish between coltan from different mines within the DRC. For example, coltan illegally mined from a deposit that extended into the Kahuzi Biéga National Park cannot be distinguished from coltan legally obtained just outside the park. This is not surprising, given that the material comes from the same geological province, the limits of which bear no relation to the park boundary.

Initiatives focused on international trade

Enough Project (Center for American Progress)
The highest profile and most extensive initiative focused on coltan is the conflict minerals campaign organized by the Enough Project. The Enough Project is funded by the Center for American Progress, a think-tank based in Washington DC with an agenda of promoting social and political change. The Enough Project focuses on crises in Sudan, Chad, eastern DRC, Northern Uganda and Somalia, and its conflict minerals campaign was launched in April 2009 to coincide with the introduction of conflict minerals legislation into the US Senate. The primary goal of the campaign is to end violence by armed groups in the Congo through better regulation and restriction of the trade in coltan, tin, tungsten and gold extracted from conflict zones. It seeks to do this through a combination of education (of governments, consumers and companies about atrocities committed by armed groups) and pressure (by shaming corporations that trade in metals originating in eastern DRC and governments that allow this trade). Another element of the campaign is persuading consumers to boycott products from electronics corporations using these metals.

The concrete objectives of the project, all focused on an overall goal of reforming conflict mineral supply chains, are: creation of a transparent supply chain, securing mining sites (using MONUC or DRC government forces), improved governance relating to mining and trade, and better livelihood options for miners. The campaign recognizes that some Congolese depend on coltan mining for a living, but argues that the best way to enable these communities to earn a decent livelihood and live in peace is to end the war. In order to do this, it argues, cutting off sources of funding to armed groups needs to be the priority.

Campaign representatives have directly lobbied politicians,

governments and corporations, but the campaign also involves signing up individuals to support its activities. Most publicity material is aimed at students and young people in the United States. Supporters have been asked to: lobby politicians to support laws that would limit the importation of conflict minerals; commit to buying 'conflict-free' electronics; lobby leading electronics firms to make their products conflict-free by sending emails stating the supporter will buy their products if the companies 'take conflict out of their products'[3]; urge their campus, schools and other institutions to buy conflict-free electronics; donate money to help advocacy efforts and fund projects in the DRC; and recycle old electronics to reduce demand for more conflict minerals. Supporters have also picketed the opening of a new Apple Inc. store to draw attention to Western corporations' use of minerals from conflict zones in the Congo. The campaign asks electronics firms to sign a pledge that they will trace the minerals they use back to the mines of origin to determine if those minerals are from sites controlled by armed groups, and 'audit their supply chains, to determine whether their minerals have been illegally taxed by armed groups'. It also asks that companies signing the pledge obtain independent verification that they are really doing what they promise. To complement the activist work of the conflict minerals campaign, the Enough Project also established the Raise Hope for Congo initiative, which aims to help Congolese women by providing funding for projects implemented by local Congolese organizations. Raise Hope for Congo has many coalition partners including domestic Congolese NGOs based in North and South Kivu, Amnesty International, Oxfam and religious organizations from the United States and Europe. Its spokeswoman is Emmanuelle Chriqui, a Hollywood actress.

The conflict minerals campaign's tactics have emphasized the use of multimedia communications, celebrity spokeswomen and engaging young people. The emphasis on youth appears to be based on their importance as consumers of electronic devices and their enthusiasm for action. The campaign website includes options for Twitter, sends regular electronic newsletters and invites budding activists to text their details and support. The Enough Project website has links to clips on YouTube that feature actors and celebrities discussing the Congo, sexual violence and conflict minerals. To raise the profile of the conflict minerals campaign and the Raise Hope for Congo initiatives, the Enough Project ran a competition for short video clips about conflict minerals. Over 100 clips were submitted and the competition was judged by a panel of celebrities. Some Hollywood actors made and submitted their own clips. At the time of writing the videos had collectively received nearly 300,000 hits.

The conflict minerals campaign is a broad-based campaign cognisant of the complex relationships between conflict, mining in war zones, development and governance. It is realistic about the integrated approach required to address these issues, including the importance of action at the international, regional and local levels. However, in publicity material these complexities are dropped in favour of simplistic messages – persuasive and skillful messages – linking consumers to the perpetuation of far-away conflicts and emphasizing the ease and urgency of legislative reform and consumer action.

United States government conflict minerals legislation
On 21 July 2010, US President Barack Obama signed into law a new Dodd-Frank Wall Street Reform and Consumer Protection Act (H.R. 4173). Primarily aimed at reducing the risk of another global financial crisis by legislating changes to the US banking and finance system, it includes a provision designed to improve transparency and reduce trade in conflict minerals from the Congo. Buried deep in the 2,300

pages of the legislation, under Miscellaneous Provisions, is section 1502 on 'conflict minerals'. The provision requires US companies to submit an annual report to the Securities and Exchange Commission that (1) discloses whether their products contain 'columbite-tantalite (coltan), cassiterite [tin], gold, wolframite [tungsten] or their derivatives' from the Congo or adjoining countries; and (2) describes measures being taken on the source and chain of custody of such minerals. Reports must describe the 'facilities used to process the conflict minerals, the country of origin of the conflict minerals, and the efforts to determine the mine or location of origin with the greatest possible specificity' [sec. 1502(b)], as well as name the entity that conducted an independent private sector audit of the due diligence measures adopted by the firm. There are no penalties imposed on companies that report they have taken no action. However, companies must make all disclosures publicly on their websites. Under the legislation the US Agency for International Development must develop a strategy to address linkages between conflict minerals and armed groups. The legislation creates a product labelling system whereby a company that cannot verify certain products do not contain conflict minerals from the Congo must describe these products in its report as 'not DRC conflict free'.

Inclusion of sec. 1502 received bipartisan support in both the Senate and the House of Representatives, and was sponsored by Senator Samuel Brownback (Republican, KS) and Representative Jim McDermott (Democrat, WA). These two politicians collaborated closely with NGOs to identify appropriate wording and generate political and public support for the provision, including the Enough Project, Global Witness (which investigates and campaigns to prevent natural resource-related conflict and corruption), World Vision (a Christian aid and development NGO) and the Information Technology Industry Council (a Washington DC lobby group representing the interests of the information and communication technology industry).

The power of the legislation is two-fold: it forces companies to pay attention to the origin of the minerals they use, and gives consumers leverage over US firms by making available product information that may shape their consumer choices. One of the tasks of the US Agency for International Development is to describe 'punitive measures that could be taken against individuals or entities whose commercial activities are supporting armed groups and human rights violations' [sec. 1502(c)]. In other words, while the bill does not specify any penalties, it provides for there to be penalties.

The consequences of the legislation are difficult to predict. Without enforceable penalties it is not clear, beyond public naming and shaming of firms and possible consumer reaction, what incentives firms have to actually put in place measures to prevent coltan handled by armed groups from entering their supply chain. The legislation might result in corporations bypassing coltan, tin, tungsten and gold originating from the DRC altogether because of the additional paperwork and cost involved in buying products containing Congolese minerals. This would be bad news for civilian Congolese producers and good news for producers from other countries, such as Australia and Brazil. If corporations comply with the legislation and post clear and correct information on their websites, this will assist concerned consumers to make an informed choice about what they are buying.

There are two wording issues that illuminate weaknesses that could undermine the intent of the provision. First, the provision can be revised or waived if this is 'in the national security interest of the United States' [sec. 1502(b)] – a not uncommon clause in legislation. However, tantalum has been routinely described as critical to hi-tech economies and in short supply worldwide. In 2000 when the price spike

occurred and there was a perceived shortage, an entrepre-
neurial politician or corporation could have easily argued that
it was in the national security interest to abandon any reporting system that made it more difficult for US companies to
obtain supplies of tantalum. Yet it was precisely at such a
moment that armed groups in the Congo were earning the
most they ever did from coltan. Under sec. 1502, it is during
those times when conflict minerals legislation could have the
most impact in the Congo that it is easiest for the provision to
be waived or revised. Second, the provision refers to conflict
minerals that 'directly or indirectly finance or benefit *armed
groups*', raising the question of what constitutes an armed
group. To find a definition one is referred to the Department
of State's Country Reports on Human Rights Practices on the
DRC. These reports distinguish between government 'security forces' and 'armed groups operating outside government
control', making it clear that – unlike this volume – the definition of 'armed group' *excludes* the DRC army. Let us recap:
members of the DRC army attack, rape and kill civilians; the
2009 Country Report even describes some of these incidents.
As discussed in chapter 3, the DRC army is also involved in
the production and trade of coltan. With the introduction of
sec. 1502, it is not clear why an American corporation would
want to buy tantalum derived from coltan, given the additional
red tape involved in doing so. However, if a firm *did* want to,
it would have a strong incentive to buy tantalum derived from
coltan that had been produced or handled by the DRC army
and it could label products made from such minerals as 'DRC
conflict free'.

Initiatives focused on demand

Several initiatives have focused on educating consumers and
encouraging them to choose certain products over others as

a way of bringing attention to violence in the Congo and the relationship to natural resources. 'No blood on my mobile', or 'pas de sang sur mon GSM' in French and 'geen bloed aan mijn GSM' in Flemish[4] (see figure 4.1) was the first campaign to specifically focus on the relationship between coltan and war in the Congo. Galvanized by the outrage that ensued after the release of the United Nations' first report in April 2001, eighteen Belgian development, human rights and church NGOs launched a campaign in June 2001 to draw attention to the role of European, especially Belgian, companies in the trade in minerals extracted from the DRC's conflict zones. In a fateful decision for the telecommunications industry, the NGOs singled out coltan from other minerals mentioned in the United Nations report because of its link to a commonplace object, the mobile phone. The campaign called on consumers and corporations to boycott products containing tantalum produced from coltan, specifically mobile phones, and implied that most mobile phones contained tantalum sourced from the Congo. The campaign asked supporters to text the message 'No blood on my mobile! Stop the plundering in Congo' to friends and to electronics corporations. A campaign in 2008 by a Dutch Christian youth organization used similar messages. The campaign, 'Break the connection', advocated for the 'sustainable and just production of cellphones'. In youth groups, student organizations, churches and at festivals, campaign representatives educated people about 'the connection between injustice and our luxury items'. Publicity material linked the exploitation of coltan and other minerals in the Congo to underdevelopment, civil war, pollution and child labour. The public was asked to write a personal message on a postcard and these were offered in a bundle to corporate representatives at a Washington DC meeting of the phone and electronics industry.

Another educational initiative, 'Breaking the Silence', is

Source: Poster reprinted with permission from Broederlijken Delen, the Lenten campaign of the Flemish Catholic Church in Belgium.

Figure 4.1 No Blood On My Mobile poster.

organized by Friends of the Congo, a Washington DC-based non-profit established in 2004. Friends of the Congo supports and funds a variety of small local development projects in the Congo, including education, literacy and rural livelihood projects. Its overall goal is to help bring peace and change to the DRC. One project involves a partnership with the Association of Widows in South Kivu to support women and children who have suffered sexual violence during the conflict, through health, education and skills development projects. Probably because of sensitivities around the dominant and damaging historical role that white people have played in African affairs, the Friends of the Congo website makes a point of stating that it is 'led by people of African ancestry and others of goodwill' and that it was established at the 'behest of Congolese human rights and grassroots institutions'.[5] A range of activities are used to raise international awareness about the Congo. There is an annual Breaking the Silence Congo Week (see figure 4.2) that features films, lectures and demonstrations about violence and poverty in the Congo, as well as mobilizing support for projects in the DRC. Congo Week events in 2008 were held on 150 college campuses in thirty-five countries. As part of Congo Week, Friends of the Congo raises awareness of the link between violence and the exploitation of natural resources in the DRC. One initiative, The Cell Out, focuses on coltan and mobile phones. The Cell Out is a coordinated boycott of cell or mobile phones from noon to one o'clock on a selected day during Congo Week. Participants are encouraged to adopt a voicemail message educating people about the Congo wars and the phone boycott. Activities are aimed at grassroots activists and students, especially from North America. Unlike some other initiatives, Friends of the Congo has not used celebrities to attract media attention, does less advocacy work among politicians and is less focused on reform of government and international trade. The organization's events,

Source: Poster reprinted with permission from Friends of the Congo.

Figure 4.2 Breaking The Silence, Congo Week poster.

grassroots membership and partner organizations resemble classic grassroots network activism.

A fourth initiative that targets consumers, this time using a message primarily about the ecological consequences of coltan mining, is the 'They're calling on you' campaign organized by Australian zoos and the Jane Goodall Institute. Launched in October 2008, the campaign's objective is to help reduce deforestation of gorilla habitat caused by mining for coltan and other minerals. A campaign poster states 'You can help save Gorillas in Africa simply by donating your mobile phone!'[6] (see figure 4.3). The campaign asks the public to donate their mobile phones, which are then either resold or broken down so the materials can be recycled in order to reduce the need for more coltan mining. The campaign to save gorillas is also linked to reducing environmental waste, stating that by donating telephones people are 'diverting [their] phone from landfill'. Funds raised from selling reuseable old phones help to support the work of the Jane Goodall Institute. While the campaign is very much focused on preserving gorilla habitat (Melbourne Zoo is a specialist gorilla centre), the campaign website also mentions the relationship between mining and ongoing conflict, stating: 'The mining of coltan within the Congo River Basin is contributing to forest loss and unrest in the region'. As part of its initiative focused on primates and the Congo, the zoo established a relationship with national parks in eastern DRC. An Australian zookeeper worked as a volunteer in a primate orphanage and reported on her experiences via a blog, and Melbourne Zoo helped pay the wages of a Congolese parks ranger.

The causal link between recycling mobile phones in Australia and saving gorillas in eastern DRC through reducing coltan mining is neither direct nor close. However, in terms of marketing, 'They're calling on you' is effective. It is tightly focused on a charismatic animal, has a simple message and

 the Jane Goodall Institute

Source: Poster reprinted with permission from Melbourne Zoo.

Figure 4.3 They're Calling On You poster.

asks people to do an easy task. Almost all Australians have a mobile phone and when they want to replace it, their old phone has virtually no value as there is no market in Australia for old mobile phones. Thus the campaign asks the public to donate something they no longer want and do not value, in order to help save an endangered species. Who could refuse that? While the campaign's educational message linking consumer choices to conservation issues is persuasive, its actual contribution to reducing demand for coltan is impossible to measure and probably minimal. The campaign also raises the question of whether recycling old mobile phones is the best way to reduce illegal coltan mining. Perhaps a better solution would be for consumers to extend the life of their old mobile phones for as long as possible and simply resist the latest Nokia or iPhone, but this is a much tougher message to sell as a way to save gorilla habitat. The campaign has also been criticized by the Australian mobile phone industry association. It argues that phones resold through the programme simply get shipped to developing countries – extending their life, but the phones still end up in landfill, only in Asia or Africa, not Australia.

Global resource initiatives relevant to the coltan industry

In addition to advocacy specifically focused on coltan, there are several initiatives designed to change the way natural resources in general are extracted and traded around the world. These global initiatives do not specifically refer to coltan or tantalum and were not created with these in mind. Nonetheless, they are analysed in order to assess whether they could complement coltan-specific initiatives in addressing a key issue at the core of coltan politics: how to prevent armed groups from profiting from coltan production and trade.

There are at least a dozen global initiatives focused on

improving aspects of natural resource extraction, trade and revenue management. Only those with concrete compliance mechanisms for accreditation and verification – essential for any scheme that aims to restructure the tantalum supply chain, especially in the DRC – are included here. Initiatives that lack compliance mechanisms and focus more on promoting best practice in environmental standards, community consultation or investment of resource revenues – such as the Natural Resources Charter and the Framework for Responsible Mining – are not discussed.

Kimberley Process Certification Scheme
The global initiative most relevant to the tantalum trade is the Kimberley Process Certification Scheme for Rough Diamonds ('Kimberley Process'), which entered into force in 2003. While it is specifically focused on diamonds, some activists, NGOs and governments consider it to be a model for supply chain management of natural resources. The Kimberley Process originated from a meeting between the governments of South Africa, Botswana and Namibia – all politically stable, diamond-exporting countries – in the city of Kimberley to discuss ways to prevent diamonds from conflict zones from entering the global diamond supply chain. The meeting was called in response to activist campaigns alleging that De Beers, historically the dominant global producer, buyer and marketer of rough diamonds, was fuelling wars in Sierra Leone, Liberia, Angola and the DRC by buying diamonds from those countries. Similar to activist arguments today regarding coltan, activists had argued that buying diamonds from armed groups engaged in war enabled those groups to earn profits and continue fighting. But activist campaigns targeting De Beers and 'blood diamonds' threatened to reduce diamond sales from all sources, and governments joined the Kimberley Process in an effort to distinguish their

diamonds from 'blood diamonds' in order to maintain sales, foreign exchange earnings and jobs. The Kimberley Process is designed to regulate the trade in diamonds by governments giving each diamond a certificate at origin and using the certificate to trace and prove its provenance all the way through to retail sale of the diamond to a consumer. It involves governments adopting a set of internal controls and regulating their own adherence to those controls.

There are important lessons from the Kimberley Process for any coltan certification scheme. It has successfully formalized the trade in rough diamonds by creating an incentive for producers to declare their diamonds in order to obtain certificates, which in turn facilitate export. An expansion of the formal diamond trade has enabled governments to levy taxes and resulted in some improvements in labour and environmental conditions. A certification scheme for coltan has the potential to create similar positive effects.

However, the Kimberley Process has not been entirely successful in realizing its initial vision. The most important criticism of the Kimberley Process is that it has not resulted in the removal of illegal actors and armed groups from the production and trade in diamonds, and it is not even certain that certification systems imposed by governments will ever be able to do this for any commodity that is valuable in small quantities and produced in widespread locations. Yet, control over extraction and trade in coltan is precisely what is most required in the DRC to stop armed groups profiting from coltan. If the Kimberley Process cannot do this for diamonds, it is unclear that any certification scheme will be able to do this for coltan. Second, certification schemes can be corrupted through both fake certificates and false certification using genuine certificates, although this does add to the cost of getting a diamond to market. For example, in 2004, a $100 million smuggling racket involving the DRC's neighbour, the Republic of Congo (Congo-Brazzaville), was exposed. Diamonds from the DRC were being illegally imported into the Republic of Congo, and then falsely certified by corrupt government officials as being from that country. Given that the Kimberley Process is run by governments and relies on governments' transparency and honesty, actions by governments to corrupt the scheme may be difficult to detect and investigate. A certification scheme for coltan will require certificates that are difficult to counterfeit, as well as controls that prevent governments using genuine certificates to falsely certify that coltan originating in the DRC was produced domestically. Third, the physical form of a diamond stays the same from when it is mined through to its final use in, for example, an engagement ring. The diamond remains in a solid state and is simply cut and polished. By contrast, tantalite is dissolved in chemicals during processing, converted back into a solid state (such as a powder or wire), and then may be mixed with another metal to create an alloy. The original ore is unidentifiable in the final product, making tantalite impossible to trace chemically or structurally once it has been dissolved. Finally, diamonds are certified as 'conflict free' when they are produced in mines not controlled by illegal armed groups (that is, armed groups opposed to the recognized state authorities). This has resulted in some perverse outcomes. For example, in 2009 Zimbabwean diamonds were certified as 'conflict free', despite reports that the Zimbabwean army killed hundreds of people when it took control of an artisanal diamond mine. Would coltan produced in mines controlled by the DRC army receive certification, especially when there is evidence of violence by DRC army troops against civilians?

Guidelines for Multinational Enterprises
Launched in 2001, the *Guidelines for Multinational Enterprises* ('Guidelines') of the Organisation for Economic Co-operation

and Development (OECD) provide voluntary principles and standards for responsible business conduct in a range of areas including employment and industrial relations, human rights, environment, information disclosure, combating bribery, consumer interests, science and technology, competition and taxation. Like the Kimberley Process, the OECD Guidelines lack any dedicated monitoring mechanism.

The Guidelines are relevant to the mining sector, but there are constraints on their relevance to mining in the DRC. Most notably, they are not applicable to the many companies operating in the DRC from countries outside the OECD that were explicitly identified by the United Nations as being involved in the illegal coltan trade, such as those from the DRC itself, China, Hong Kong, Malaysia and South Africa. One advantage of the Guidelines is that they apply to any OECD multinational enterprise, not just mining firms, allowing coltan processors such as Cabot and H.C. Starck to be monitored. Crucially for the coltan industry, however, the Guidelines have little to say about smaller suppliers to multinational corporations, other than that suppliers are encouraged to operate in accordance with principles laid out in the Guidelines. The Guidelines are also simply that: guidelines regarding a set of principles that corporations may voluntarily choose to follow. OECD governments have established 'national contact points' to deal with breaches of the Guidelines, but contact points rely on interested third parties, such as NGOs, to investigate breaches and develop a case against the firm. Investigations result in recommendations, the implementation of which is not monitored by any government authority.

Despite these innate weaknesses, the Guidelines have had an impact and their power lies in their usefulness in shaming governments and corporations that care about their public reputation. OECD governments were embarrassed by the United Nations' findings that they failed to fulfil their monitoring obligations under the Guidelines because they did not keep track of or take action against their corporations that were involved in the illegal trade in natural resources from eastern DRC. The United Nations reports of breaches of the Guidelines were the catalyst for subsequent parliamentary inquiries held by the Belgian, British and French governments into the activities of corporations in the DRC. Global Witness, an NGO, also successfully used the Guidelines to pressure the British government into investigating a British-registered company named in the report (see chapter 5).

Global Compact
The Global Compact, an initiative of the UN launched in 2000, involves individual businesses agreeing to ten principles in the area of human rights, labour, environment and corruption prevention. It claims to be the largest corporate citizenship and sustainability initiative, with more than 7,700 participants, including over 5,300 businesses in 130 countries. The Global Compact includes some well-known transnational corporations and many small and medium-sized enterprises. One hundred and seventy-one companies are listed in the Industrial Minerals and Mining Sector, including many of the largest players. Signatories to the Compact are supposed to ensure their operations and strategies accord with each of the principles. The principles are not legally binding and adherence is voluntary, but non-compliance by signatories results in them being delisted.

Like breaches of the OECD's Guidelines, being delisted from the Global Compact may cause embarrassment to a corporation that cares about what ordinary citizens think and may precipitate changes in corporate practices. However, publicly owned corporations are probably more vulnerable in this regard. Corporations that are privately or state-owned, or which rely on joint ventures with governments, are less susceptible to public shaming unless this threatens their operations.

The advantage of the Global Compact is that it functions as a baseline in terms of standards, as all its principles are linked to United Nations declarations that have been nominally accepted and endorsed by UN member states. However, the principles are cumbersomely worded and often only make sense when read in conjunction with the declarations. Some principles are at odds with national laws, such as those concerning labour unions. The principles are most suited to transnational corporations that deal with many different governments and can use them as a common reference point. However, full compliance would be difficult to obtain and onerous to monitor for corporations operating scores of mines worldwide. Firms that work in a single country are more likely to care about the laws of that country than about universal declarations.

The Global Compact has some relevance to coltan production and trade in that signatories should desist from using violence against labour and operate within the law. However, its usefulness in the illegal and semi-legal grey zone of artisanal and small-scale coltan mining is minimal, especially as the DRC government cannot even enforce its own mining code. Congolese are also likely to be sceptical, and possibly hostile, towards *any* initiative run by the United Nations, given its failure to end violence in the DRC. At best, minerals processing firms outside the DRC that are members of the Global Compact might refuse to buy ore with a provenance that potentially originates in the DRC.

Extractive Industries Transparency Initiative
The Extractive Industries Transparency Initiative (EITI) was launched in 2002. It was created out of concern that profits from oil were not being used to benefit the citizens of the

countries in which oil is extracted, because either profits were being embezzled and misspent by governments, or repatriated by foreign corporations. The EITI brought together governments, companies and civil society organizations to establish benchmarks relating to (1) the transparency of information about the commercial arrangements used to extract oil, gas and minerals, and (2) independent verification and certification of the management of revenue from natural resources. Countries that achieve the benchmarks have their compliance status validated through third-party checks. The first country to achieve compliance was Azerbaijan in February 2009. In 2010, twenty-seven countries, including the DRC, were candidates for validation. If a government signs up to the EITI, data on how it collects and manages oil revenue is made public. The purpose of improved transparency is to increase governments' accountability to citizens, and to make corruption more difficult and easier to detect when it does occur.

The EITI's certification system could potentially be adapted to resources other than oil, such as a country's tantalite mining industry. However, *countries* receive validation, not specific resources. Ultimately, the EITI depends upon voluntary changes by governments to domestic laws to enforce transparency of information about revenue collection from the extractive sector. The DRC could conceivably implement and achieve compliance with an EITI-like system for minerals such as copper, which is exploited in a controlled manner in Katanga province. However, compliance for copper would not make the DRC eligible for certification, as the exploitation of other minerals in eastern DRC would not achieve the benchmarks.

There are three criticisms of the EITI that are relevant to coltan politics. First, it is voluntary; the countries that most need to bring transparency to oil revenues because of widespread corruption, are also most likely to not sign up. For example, Angola has decided not to join, although as mentioned the DRC is a candidate for validation. Second, the EITI focuses on resource revenues, rather than *expenditure* of revenues. That is, while profits from natural resources are tracked through to a country's treasury, what happens to those revenues afterwards remains opaque. Clearly, combating corruption and increasing a government's accountability to its citizens must also involve transparency in how revenues are spent. Third, the system of using multiple stakeholders to validate a country's compliance runs the risk of 'regulatory capture', whereby the government, oil companies and civil society participants identify with the country undergoing the validation process and permit validation for reasons of the 'national good'. Stakeholders may also approve validation because they have been bribed or offered favours. Civil society may also not be as independent as one would expect: while independent civil society can be a trenchant critic of government, political and business elites sometimes create civil society organizations to achieve political and economic ends. Such organizations could become the 'official' civil society stakeholder in an EITI process.

For the purposes of this volume, the key test of the EITI is whether it could effect changes in the way armed groups currently earn profits from coltan. In regard to FDLR, Mai Mai and other anti-government militias, clearly it could not because these armed groups do not obey the law anyway and stand to gain nothing from increased transparency. The armed group for which the EITI could theoretically have a transforming effect is the DRC army. If the government received certification under the EITI and actually enforced new requirements for transparency in the natural resources sector, involvement by the DRC army and individual Congolese politicians in coltan production and trade would be exposed. Any resulting international and domestic outcry could result in the army and political figures withdrawing from coltan mining.

Publish What You Pay

The Publish What You Pay campaign was launched in 2002 by a coalition of mostly British development and human rights NGOs concerned at the mismanagement and embezzlement of revenue from oil, gas and mining extracted in developing countries. Publish What You Pay works through local partner organizations in resource-rich countries to demand transparent and accountable management and expenditure of public funds to address poverty and corruption. It also aims to reform autocratic government by facilitating public participation in debates over revenue management. The campaign involves local civil society groups in seventy countries. Publish What You Pay calls for the mandatory disclosure of extractive industry contracts and for licensing procedures to be carried out transparently in line with best international practice. The campaign focuses on all minerals, although at the time of writing there were no specific actions on tantalum. It has an office and a dozen local partner organizations in the DRC.

The strength of Publish What You Pay is that it uses a third party – civil society – to monitor corporate and government actions. Its weakness is that if civil society is disorganized, lacks funding or cannot engage with revenue management issues, minimal leverage can be brought to bear on governments or corporations. The DRC has a robust civil society, indeed the health and education systems are largely run by non-profit organizations. But civil society is organized around compensating for the weaknesses of the state, not challenging it on policy issues. To be effective, local Congolese NGOs that pursued a Publish What You Pay agenda would need external funding and nerves of steel in facing down a government accustomed to using the state security apparatus to harass civilian critics.

Sustainable Development Framework
The International Council on Minerals and Mining ('Council') is an umbrella organization of some of the most powerful transnational mining and metals corporations in the world, none of which have a current interest in coltan. Following its participation in a 'Mining, Minerals and Sustainable Development Project' in the early 2000s that researched how mining could be better balanced with environmental and social concerns and the development of transparent governance systems for revenue, the Council developed a Sustainable Development Framework ('Framework') for the minerals and mining sector.

The Framework has three elements: a set of ten principles, public reporting requirement and an independent assurance mechanism. In contrast to the principles of the UN Global Compact, the ten principles focus explicitly on the business practices and decision-making processes of the mining and metals industries. Nevertheless, the intent and impact of the Council's principles, if applied, would be similar to that of the Global Compact. If firms that buy and trade coltan adopted the principles, they would have to put pressure on the DRC government and producers to implement measures to ensure armed groups do not profit from the coltan supply chain, or they would have to buy tantalite from other sources. The principles thus have the potential to change the behaviour of firms in the global coltan supply chain, but mainly through internal corporate reform. They do not include any role for government or civil society to apply external pressure on companies, and compliance is voluntary.

Initiative for Responsible Mining Assurance
The Initiative for Responsible Mining Assurance ('Assurance Initiative') was launched in Vancouver in 2006. It is supported by some major mining corporations, jewellery retailers, trade associations and non-profit environmental organizations. The

Assurance Initiative was created to develop and establish a voluntary third-party system to independently verify compliance with environmental, human rights and social standards for mining operations. The system enables retailers, such as jewellers, to claim that their products contain minerals from mines meeting best practice environmental, human rights and social standards. The attraction for retailers, such as Tiffany & Co., is that it may be able to create a market for 'ethically produced' gold and diamond jewellery, and even charge more for these products. This would be a win for retailers, as well as for mining communities if they receive a share of the increased profits. The Assurance Initiative is different to the EITI and Publish What You Pay campaigns in that it does not focus on transparency in revenue management; rather, it emphasizes the conditions in which mining occurs. The Assurance Initiative covers tantalite mining, but it is more applicable for large-scale mining by corporations than the kind of artisanal and small-scale mining that produces coltan at myriad sites in the DRC.

Table 4.2 lists the agents that the global initiatives envisage will bring about change in the resources sector. It also illuminates the challenges global initiatives would face in separating armed groups from coltan production and trade in the DRC. Global initiatives that rely on cooperation from mining companies and strong state institutions in the country in which mining occurs are unlikely to have an impact on the coltan supply chain, given the weakness of the DRC state, particularly its inability to enforce mining, labour and environmental laws in eastern DRC. Initiatives that rely on civil society obtaining and releasing information about revenue management, and/or providing advice to mining companies, are also unlikely to have much impact. While Congolese have been angry for 130 years at the way their rulers spend resource revenues, Congolese civil society faces tight government controls over economic

Table 4.2 Global minerals initiatives: key agents for change

	Key agents for change in restructuring resource supply chain						
	Extractive corporation	Government at extraction site	Government at corporate HQ	Processors and manufacturers	Retailers	Consumers	Civil society*
Kimberley Process	✓	✓		✓	✓	✓	
OECD Guidelines	✓		✓		✓	✓	
Global Compact	✓			✓			
Extractive Industries Transparency Initiative		✓					
Publish What You Pay	✓	✓					✓
Sustainable Development Framework	✓			✓			
Initiative for Responsible Mining Assurance	✓				✓		✓

Source: compiled by author using information from initiatives.

*Civil society's role in these initiatives is to publish information and/or provide expert advice to businesses and communities.

information and there is no government tradition of consultative economic policy-making. Finally, due to the absence of large transnational mining corporations in eastern DRC, initiatives such as the Sustainable Development Framework and the Initiative for Responsible Mining Assurance, that rely on companies adopting certain principles, will not have much of an impact – at least, not until transnational corporations invest once again in eastern DRC. A key weakness in all these global initiatives, with the exception of the Kimberley Process, is that their approach to natural resources is on a country-by-country basis, whereas a *supply chain approach* is required to prevent the entry into the international market of coltan handled by armed groups. The global initiatives likely to have greatest impact on coltan production and trade are those that rely more on change agents outside the DRC, that is, a modified version of the Kimberley Process and the OECD Guidelines (if actually enforced by OECD governments). The Sustainable Development Framework and Initiative for Responsible Mining Assurance could also possibly have some effect through members that are minerals processors or retailers that stock electronic devices containing tantalum.

Understanding coltan advocacy

Advocacy centred on coltan is part of a tradition of activism focused on Africa and the Congo. The Congo has been central to some of the world's landmark activist campaigns, beginning with the campaign in the eighteenth century to abolish the Atlantic slave trade and then to outlaw slavery, which occurred in the British Empire in 1838. Sixty years later the Congo, then the private possession of King Leopold II of Belgium, became the focus of the twentieth century's first great transnational human rights campaign: to end Belgians' brutal exploitation of Congolese slave labour to extract rubber and ivory. More broadly, for over 200 years activists and the international community have engaged in campaigns to improve the welfare and rights of Africans. In addition to campaigns against the slave trade and abuse under King Leopold II's regime, activist campaigns concerned with Africa have focused on apartheid, the Rhodesian government, famine in Ethiopia, blood diamonds, indebted governments, and violence in Darfur, Sudan. The premise of these campaigns was that action abroad, especially in Western countries, could change conditions in the country in question.

Campaigns by NGOs against coltan are not new in their focus on a natural resource. Perhaps the most well-known campaign about a natural resource was the campaign against blood diamonds. This campaign contributed to the establishment of the Kimberley Process, and advocates of more regulation in the coltan industry directly draw on it for inspiration and ideas. Other activist campaigns against natural resources target 'dirty gold' (produced using mercury or cyanide which is then released into the environment), 'conflict cocoa' (used to fund armed groups waging civil war in Côte d'Ivoire), and logging in the Amazon and Borneo. Related initiatives focus on the labour conditions under which natural resources are produced, such as 'fair trade' campaigns that certify a product as having been produced under fair labour conditions, with minimal environmental damage and an adequate return to the local producer. Fair trade products, including coffee, cotton, timber and chocolate, have found a niche retail market despite being more expensive than uncertified equivalents.

Structure and organization of coltan initiatives
What is the best way to understand the structure of, and relationships between, disparate coltan initiatives conducted by separate organizations, all focusing on some kind of reform of the coltan supply trade but lacking strategic coordination

between various groups? Are they a movement? A coalition? A network? *Movements* are 'sustained interactions between challengers and authorities on matters of policy and/or culture.'[7] Different parts of movements may not engage with each other or even be aware of the existence of other parts, but there is a similar shared goal of policy or cultural change. Examples of movements include the US civil rights movement of the 1960s, the feminist movement of the 1960s and 1970s and the movement against the US-led invasion of Iraq in 2003. *Coalitions* involve different organizations pooling their resources and cooperating to take advantage of similar objectives in order to achieve shared goals. Often this involves the joint planning of activities in order to take advantage of opportunities. Coalitions can be long-lasting or short-lived, often dissolving when the goal has been achieved or fails. Examples of coalitions include: collaboration between environmental organizations and anti-nuclear organizations to stop uranium mining in Australia in the 1970s and 1980s; collaboration between churches, labour unions and the main commercial farmers' organization in opposition to the Mugabe government in Zimbabwe in the late 1990s and early 2000s; and collaboration between Oxfam America, Earth Works Action (a Washington, DC-based mining activist organization), CooperAcción (a Peruvian activist organization) and local farmers to stop new gold mines in Peru in the 2000s.

Coltan campaigns have movement-style interaction between activist organizations and foreign governments to change policy, specifically to restrict the trade of coltan. Foreign governments and DRC government authorities at both the national and provincial levels are involved in some initiatives to reshape the coltan industry. However, there is no useful interaction with the 'authorities' in the conflict zones where coltan is mined, that is, interaction with armed groups such as the FDLR and the DRC army. Coltan campaigns also lack the kind of mass public engagement and protest that occurred in the feminist, civil rights and anti-war movements. While coltan campaigns do not constitute a movement, some do exhibit the collaboration typifying *coalitions*. For example, the 'no blood on my mobile' campaign was organized by eighteen different NGOs; the Durban Process for Ethical Mining brought together environmental organizations, private charities, corporations, the World Bank and the United Nations; and the Enough Project collaborated with United States politicians and other NGOs on conflict minerals legislation. However, there is little collaboration *across* campaigns, which are organized and planned unilaterally and do not share resources. Coltan campaigns can best be described as a network. There is diffusion of information and ideas between different organizations with an interest in coltan, but there is little strategic cooperation to jointly plan action to exploit opportunities. Collaboration between US partitions and NGOs on conflict minerals legislation is a significant exception.

Congolese attitudes to coltan campaigns
Congolese attitudes to coltan campaigns will vary according to the relationship of the individual being asked to the supply chain. However, it is difficult to say what Congolese attitudes to coltan initiatives are because no one appears to have asked, at least, not in any comprehensive way. It is also logistically difficult to survey the opinions of a statistically significant sample of Congolese, especially far-flung *creuseurs* and *négociants* most exposed to pressure from armed groups.

Television documentaries have captured some Congolese views about suggestions that Congolese tantalum be banned – although this is not, in fact, what campaigns demand. Most campaigns argue that illegally mined coltan should be kept out of the supply chain while a trade in legally mined ore should be encouraged, but these nuances do not get captured on film.

One coltan trader in South Kivu, when asked for his thoughts on the 'movement' in the West to stop buying coltan, replied

> This is another way of penalizing Africans, in particular, but also the South Kivutians and the Congo. We have difficulties and I think if there was a small injection of money into mining activities it would be enough – instead of continually giving us aid that leads us nowhere. You people outside say coltan this, coltan that, but what is the truth? . . . Our life depends on this.[8]

A representative of a *comptoir* interviewed by Global Witness had a similar opinion: 'We all end up buying minerals which, in some way, have been produced illegally. You can't just ask us to stop. We have no alternatives other than closing.'[9]

As noted, Congolese opinion is varied. Some Congolese are against a ban on coltan, but are in favour of phone boycotts as a way of drawing attention to conditions in the DRC. The president of one expatriate community association argued that a mobile phone boycott 'would raise awareness of the situation in Congo . . . get people to think about coltan and human rights in Congo'.[10] Another Congolese, a former coltan trader, disagreed. When asked what he thought of arguments that coltan was a cause of war and should therefore be boycotted, he replied 'No, no, no. In my opinion [armed groups] not only get money from coltan, but from gold and other minerals. A boycott is not the solution because the war is not brought by the existence of coltan. Coltan is not the reason for war.'[11]

Challenges facing coltan campaigns

Changing government policy, global trade patterns, consumer habits, property rights and corporate purchasing habits is difficult, but this is what initiatives aim to do for coltan and conflict in the DRC. It has not been easy and will not be straightforward. There are major marketing, logistical and tactical challenges in getting people, corporations, governments and multilateral organizations to engage in coltan issues and change their behaviour or policies.

Congolese involvement

A key challenge is that there are no ready constituencies outside the DRC interested in the welfare of the Congolese. There are some Congolese diaspora communities, especially in the United States, France, Belgium and South Africa, but they are politically and ethnically divided and are not wealthy. Campaigns focusing on sexual violence in the DRC have tried to create a constituency by appealing to women. However, women are a broad constituency with multiple interests and identities that overlap with class, race, ethnicity and religion. One community that has been historically prominent in supporting development projects and raising awareness of poverty and conflict in the DRC is the Christian community (many Congolese are devout Christians). Any visitor to the DRC will quickly meet an assortment of missionaries – Catholic, Baptist, Mennonite, Anglican, Evangelical – and Christian churches and networks remain prominent advocates for Congolese issues today. However, some missionaries left during the years of conflict and there are much easier countries in which to work. Campaigns focused on coltan thus have to attract disparate groups from many countries who have different interests and values. Broader engagement with coltan campaigns has not spread far beyond Christian and student communities in the United States and Belgium. There are Christian church networks in these countries with direct person-to-person links to congregations in the DRC, and these networks are very organized. Engagement with students in the United States is largely the result of the Enough Project and Friends of the Congo. There is also residual interest by Belgians in Congolese affairs because of colonial history.

What stands out about most coltan initiatives is the lack of any substantial involvement by Congolese in the conception and organization of initiatives. However, some campaigns, including the Enough Project and Friends of the Congo, have Congolese spokespersons and supporters, and other initiatives, such as the Durban Process, consulted with Congolese about what activities should be implemented. The DRC government is involved in the initiatives that are taking place in the Congo and its cooperation is vital for these to work. However, the DRC government does not control the funding and is the weaker partner. One reason for the lack of domestic involvement by Congolese organizations is probably that within the DRC there is no broad, organized, domestic network focused on coltan issues, which would otherwise be able to act as a local partner and the eyes and ears for foreign NGOs with a coltan campaign. There are Congolese, including scholars and church leaders, who have spoken out about aspects of the coltan industry, but this is different to international campaigns having a capable and organized domestic organization to work with. Like the campaign against blood diamonds, coltan campaigns and the broader movement focused on conflict minerals lack constituents from within the affected countries. Ideas, tactics and resources come from outside.

International campaigns with a focus on coltan advertise their partnerships with local organizations in the DRC, but these are development, community or human rights organizations, not organizations comprising coltan miners, traders or affected communities. There appear to be no domestic organizations that explicitly represent the interests of coltan miners or traders, with the possible exception of collectives of coltan and cassiterite miners in Walikale. There are, however, broader organizations that represent business interests, including mining interests, across the DRC, such as the Fédération des Entreprises Congolaises. The lack of

Congolese organizations focused on the interests of coltan miners or traders is not surprising, given the transient artisanal nature of coltan mining, the conflicted conditions in which it occurs and the fact that most miners and traders handle several different minerals.

Perhaps involvement by Congolese in coltan campaigns is too much to ask, given the difficulties of language and communication between eastern DRC and the outside world. Direct involvement by Congolese in organizing campaigns is not even necessary to reform parts of the coltan supply chain, as key downstream pressure points on the supply chain lie far from Africa. However, cooperation with the DRC government and Congolese mining communities will be required at some stage, especially for initiatives focused on reform of mining and trading practices. Advocacy, campaigns and initiatives organized without Congolese input run the risk of once again making foreigners the architects of the Congo's future.

'Africa fatigue'

Another factor mitigating against coltan initiatives is 'Africa fatigue', a feeling that the continent is so hopeless and its problems so vast, despite so much energy and money already spent, that there is no point trying any more. This feeling may be more acute when it comes to the Congo. As one Australian community leader involved in refugee rights and Congolese issues summed up: 'There's a lot of compassion fatigue. A lot of people talk about Congo, but I don't see rallying because they put it in the too hard basket.'[12]

Related to both compassion fatigue and the complexity of issues surrounding coltan and the Congo War is uncertainty by potential supporters of campaigns about what action they should take. One activist, when asked whether she had changed her behaviour as a result of learning about coltan and the relationship to armed groups in the DRC, commented:

My behaviour hasn't changed and this is why I feel so frustrated . . . You want to purchase things that are ethical but you don't know what is. The coltan has gone through so many hands you don't know what is and what isn't ethical. How is the average person supposed to know?[13]

The difficulty in knowing what to do is partly about confusion around the cause-and-effect relationships between coltan, other minerals, armed groups, mobile phones and violence in the DRC, but also partly about there being so many global justice issues in which to get involved and whether involvement can make a difference. This confusion and frustration is yet another major challenge for coltan initiatives. It is easy to participate in simple initiatives such as phone boycotts or recycling. However, if and when participants start to wonder whether their actions will have a direct impact on the cause of conflict in the DRC, campaigns need to be able to provide proof of impact or their credibility will be undermined.

Publicity and marketing

Activist campaigns gain publicity if they have focal points to which the public – and even governments and multilateral organizations – can be directed and around which supporters can be organized. Coltan advocacy lacks such focal points. There has been no specific outrage (such as a massacre of *creuseurs*) or specific international meeting that activists could target and market as *the* event demanding 'Action! Now!' Violence, atrocities and peace talks in the Congo have simply been iterated in a banal continuum. By contrast, for example, the peace movement against US-led invasions of Iraq in February 1993 and February 2003 had a pending battle to protest, and anti-globalization demonstrators have regular World Trade Organization meetings to target.

One reason some advocacy campaigns stick in the public mind is because their cause is linked to an individual who comes to embody the campaign or movement. Aung San Suu Kyi fulfils this role for pro-democracy Burma activists and her house arrest was symbolic of repression in Burma. Similarly, the Dalai Lama is synonymous with Tibet independence campaigns, Nelson Mandela is forever associated with the anti-apartheid movement, and Rigoberta Menchu is the face of the Guatemalan indigenous rights movement. The images and words of these individuals play a critical role in attracting supporters and sparking the public's interest. Coltan campaigns, like other campaigns focused on a generic natural resource, lack an individual figure who can be readily identified in the popular imagination as the 'face of coltan'.

Other activist campaigns have attracted public interest because of the celebrities who become involved. Princess Diana was an advocate for the anti-personnel (land) mines campaign; Bono, the lead singer of the rock band U2, is an advocate for debt relief, international development and HIV/AIDS initiatives for Africa; Richard Gere is an advocate for Tibetan independence; Angelina Jolie is an advocate for refugees and a spokeswoman for the United Nations refugee agency, UNHCR; and George Clooney champions the cause of human rights in Darfur, Sudan. Iman, the Somali former supermodel, spoke out against blood diamonds and terminated her contract with De Beers to protest its trading in diamonds from conflicted countries. The Enough Project has attracted several American film and television stars to participate in its activities: Brooke Smith, Emmanuelle Chriqui, Julianne Moore, Julianna Margulies, Nicole Richie, Robin Wright Penn and Sandra Oh. Iman, now a cosmetics entrepreneur, has also thrown her support behind the Enough Project's campaign. The publicity that the Enough Project's high-profile supporters engage in is largely about sexual violence perpetrated by armed groups and the profits made by the latter from conflict minerals, thus the involvement of female celebrities. Nevertheless, celebrity involvement has not extended to the occupational safety or property rights of coltan miners.

Celebrity power has its limits. Famous actors can galvanize consumers to participate in product boycotts and even convince corporations to declare adherence to new principles and codes. But getting the private sector – especially trading firms near the start of the minerals supply chain – to change corporate practice is another matter. Western politicians care about celebrities because associating with them can improve their image and make government initiatives more popular. President George W. Bush spent time with Bono, who momentarily gave the President a cool image. But non-Western leaders are less susceptible to celebrity power. Presidents Joseph Kabila of the DRC, Paul Kagame of Rwanda and Yoweri Museveni of Uganda, for example, gain little from association with Western celebrities, although association with local celebrities may help their public image.

Successful advocacy needs images. Think of the photograph of the Vietnamese girl napalmed by the United States military used in anti-Vietnam War campaigns, nuclear mushroom clouds or the surviving church at Hiroshima used in the nuclear disarmament campaign, and sketches of a packed slave ship used in the anti-slavery campaign in Britain in the eighteenth century. Unfortunately for coltan activism, there are few images that readily lend themselves to coltan campaigns. Photographs of murdered civilians or victims of sexual violence do not show a causal link to coltan mining, even while they may precipitate public condemnation of the perpetrators. The coltan campaigns that are focused on gorillas make the best use of images: compelling photos of gorillas looking reproachful – as though they cannot quite understand why we are not doing enough to help. Exhausted coltan miners make for good documentary photographs, but these rarely capture the dangers and connections to violence. The lack of good 'coltan images' is probably one reason publicity material of the Enough Project emphasizes images of celebrities, and the occasional Congolese.

The image that has been commonly used is a mobile phone, which is supposed to suggest connections between Western consumers, armed groups and violence against civilians. See, for example, the posters produced for the 'no blood on my mobile' and 'breaking the silence' campaigns. Campaigns also use plays on words around phones, such as 'the cell out' and 'break the connection' to propagate the idea that the metals in mobile phones are instrumental in violence in the Congo. However, mobile phones are not the perfect image and they are no substitute for a person, let alone a gorilla. A mobile phone may or may not contain tantalum from Congo, so an image of a mobile phone is necessarily generic, making its link to the Congo uncertain. Coltan inside a mobile phone or laptop is not even visible, unlike a diamond engagement ring where the object of activism is evident on someone's finger.

Some of the rhetoric and publicity of certain coltan initiatives are effectively intellectual sleight-of-hand, portraying sexual violence in the DRC as being the direct result of armed groups trying to win control of coltan (and other) mines, or a decision to recycle a mobile phone as directly reducing demand for coltan in the Congo. Both claims are nonsense. The causal relationship is either distant or non-existent if the phone does not contain coltan, or does not contain tantalum at all. It is true that armed groups that control coltan mines also perpetuate sexual violence as, indeed, do armed groups that do not control coltan (or any) mines. It is also true that if millions of people refused to buy mobile phones with capacitors made from coltan, companies making mobile phones would probably quickly turn to capacitors containing tantalum from non-Congolese sources or use capacitors made from other metals or ceramics.

Experienced activists are aware of the limitations on what a campaign can be expected to achieve and are also alert to over-simplification. As one activist commented, 'I'm afraid [a mobile phone boycott] might be a token thing that does not make a difference. But if it just educates people and catches their attention and the media's attention then maybe it can change something.'[14] Friends of the Congo's 'breaking the silence' campaign uses phone boycotts for this purpose, as did the original Belgian 'no blood on my mobile' campaign. The Enough Project, the most extensive and sophisticated coltan campaign, treads a fine line. It proposes consumer boycotts of products unless the firms that make them can guarantee the products contain no conflict minerals, and argues that such boycotts will reduce violence in the Congo. The danger is that if violence at the local level does not subside following company-based consumer boycotts, the Enough Project's broader – and less exciting – agenda of law reform, capacity building, enhanced human security and corporate engagement could become discredited.

Conclusion

Past initiatives organized or precipitated by activists have been instrumental in changing the lives of Africans and Congolese. Will coltan advocacy, campaigns and initiatives being pursued in the 2010s make a difference? The flurry of initiatives in 2009 and 2010, including new legislation in the United States, the coltan certification scheme and the launch of the Enough Project's conflict minerals campaign, makes it easy to forget that eastern DRC has experienced conflict for two decades. This puts in context the hard slog NGOs have faced. The fact that governments and corporations are, finally, paying attention to the supply chains of minerals from eastern DRC is a victory for NGO advocacy. As John Prendergast,

co-founder of the Enough Project, commented after the US
Congress passed conflict minerals legislation in July 2010, 'A
year ago most members of Congress hadn't even heard of conflict minerals.'[15] The reason they now have is due to the efforts
of activists and NGO advocacy groups, including NGOs that
do not have specific coltan campaigns but which have persist-
ently reported on violence over the past decades: Amnesty
International, Global Witness, Human Rights Watch, the
International Peace Information Service and the International
Rescue Committee. The United Nations reports and resolutions also helped put and keep the conflict on the international
community's radar screen. Success in drawing attention to
ongoing violence does not, however, mean that the solutions
proposed by NGOs and activists will end the violence. The
issues that make coltan and other mining contested today,
such as access to land and contested property rights, remain
unresolved and largely unaddressed by NGOs, governments
and MONUC.

A major difference between the coltan-focused initiatives and the global initiatives for natural resources is the
significance they assign to consumers as agents of change.
Consumer advocacy involving boycotts is one of the key strategies of coltan initiatives, but of the global initiatives only the
Kimberley Process assigns any role to consumers. By contrast,
the global initiatives emphasize changing the upstream end
of the natural resources supply chain, as do half of the coltan-
focused initiatives. Clearly there is a dire need for changes to
the domestic production and trade of coltan, which lie at the
heart of coltan politics. However, current political and security conditions in eastern DRC mitigate against most of the
global initiatives having much of an impact – indeed, these
conditions preclude the DRC government from qualifying for
or being able to participate in most initiatives. Thus despite
the aptness of global initiatives for addressing many of the

issues activists identify as being wrong with the coltan supply chain, until conditions substantially improve in eastern DRC it is action external to the DRC that is more likely to have an impact.

If peace returns to eastern DRC – and there are fleeting signs that it gradually, if fitfully, may do so – there will be a mining boom, as occurred in Katanga after the peace accords in 2002. Development of the mining sector is a priority for the DRC government and donors, including the World Bank, which estimates that by the late 2010s mining could contribute up to 25 per cent of GDP and 33 per cent of government revenue. But any mining boom will be accompanied by changes to the sector as a new actor in African mining deftly carves out a role for itself in the DRC through joint ventures and infrastructure deals: China. Whether the achievements of coltan initiatives endure and the tactics of activists remain pertinent will in large part depend on the course of future growth and development in the DRC's minerals sector.

The ongoing relevance of consumer-focused coltan initiatives also depends on the continuing importance of Western consumers as buyers of products containing tantalum. In this respect, coltan initiatives wanting to use phone boycotts as a tactic should consider downplaying their potential impact on armed groups' profits and instead use the boycotts more as a device to generate awareness of violence in the Congo. This is because, just as the DRC mining sector is set to change, so too is the international market for electronic devices. And here, again, China and its 1.3 billion consumers – along with consumers elsewhere in Asia, Africa and Latin America – are playing a central role.

5

The future of coltan politics

The challenge for coltan activists has been to generate public interest in a war in far-away Congo and to inform people of their connection to the conflict through the products they use. The mining industry in the DRC and internationally has always been aware of the connectedness of consumers on one side of the world to producers on the other side, for minerals and metals markets are global. These connections were made abundantly clear in the late 2000s as the global financial crisis exposed the DRC's dependence on the mining sector, especially in Katanga where mining companies retrenched thousands of workers and scaled back new projects.

The financial crisis illuminated something else as well: as Western economies stagnated, there appeared to be a shift in economic power, away from OECD countries, to Asia with its buoyant economies and 4 billion consumers. Consumer boy-cotts have been a tried and tested tactic for activists trying to draw attention to the conditions under which products sold in rich countries are produced in poor countries. Boycotts of mobile phones – either on the basis of brand, or not using them for a target period – are also central to NGOs' coltan campaigns. Will such boycotts remain a useful tactic in a world where the consumers who matter speak Chinese, Hindi or Indonesian?

This chapter analyses the future of coltan politics, includ-ing demand for tantalum and coltan campaigning. It argues that coltan advocacy has brought about real changes in terms of changing corporate practices. However, due to ongoing demand for tantalum, coltan produced and traded by armed groups continues to find buyers, especially from China, which is an increasingly important manufacturing centre for products containing tantalum. The emergence of developing country markets for products containing tantalum is a manifestation of a broader shift in global consumer power. For some commodities, developed markets will remain critical and therefore the established tactics of Western-based NGOs – such as consumer boycotts and public shaming of corporations – are likely to remain useful. What the politics of coltan illuminate, however, is that if activists with transnational goals are to maintain their hard-won influence over how corporations produce and trade natural resources, they need to follow the example of transnational corporations and engage with Asia.

Industry response to coltan campaigns

Coltan campaigns have made a difference to the global supply chain. The initiatives of the United Nations and NGOs have changed the buying patterns of tantalum processors and of corporations that buy components made from tantalum to manufacture their products. The precise contribution of coltan campaigns to changing corporate practices is, however, difficult to distinguish from other initiatives focused on other minerals such as gold and cassiterite. The United Nations reports made governments aware of the role played by companies, including from Western countries, in the illegal exploitation of natural resources during the Congo War. Following the release of the first reports it was NGOs and the media that more broadly publicized the findings, and NGOs that organized campaigns to apply pressure on governments and companies. This publicity and pressure ultimately persuaded the British, Belgian and French governments to launch inquiries into the role of corporations from their countries in the illegal natural resources trade in the DRC.

Once NGO and governmental pressure was applied, corporations quickly began to take action, or at least claim that they were doing so. In 2001, Swissair and Sabena suspended transportation of minerals from Central Africa. In 2001, Cabot announced that it had cancelled all orders of tantalum from Central Africa and throughout the 2000s other companies made similar announcements that they would no longer buy coltan or use products containing coltan, including H.C. Starck, Nokia, Motorola, Apple Inc., Intel and Hewlett Packard. Researchers who have discussed conflict minerals issues with private corporations involved in extractive industries also report that the international private sector is 'increasingly uneasy about negative advocacy against the trade, which may lead to a complete disassociation from D.R. Congo, rather than a selective engagement'.[1] Some corporations also responded by engaging with NGO-sponsored coltan initiatives. Soon after the publication of the first United Nations report on the illegal exploitation of natural resources in the DRC, representatives from European telecommunications firms and the tantalum industry association, the Tantalum-Niobium International Study Center, met with the NGO, Fauna and Flora International, to discuss the impact of coltan mining in the Congo. H.C. Starck helped fund the Durban Process for Ethical Mining.

Announcements by Western firms that they no longer buy coltan are criticized by NGOs and activists as being merely words, with little real proof that the companies have stopped using coltan. But corporations are easy targets in this sense. Even if they have sound sourcing and auditing processes in place, it will always be easy for activists to claim that illicit coltan is getting into the tantalum supply chain. As with any negative claim, it is difficult for corporations to definitively prove that they are *not* doing something, but easy for activists to claim that they secretly *are* doing something.

Campaigns focused on coltan and other conflict minerals contributed to firms in the electronics and telecommunications industries deciding to act collectively on corporate social responsibility issues. Two prominent initiatives were launched: the Global e-Sustainability Initiative (e-Sustainability Initiative), by predominately European firms; and the Electronics Industry Citizenship Coalition ('Electronics Coalition'), by predominately North American firms. Some corporations are members of both. The e-Sustainability Initiative was established in 2001, around the same time as the release of the United Nations' first report and the 'no blood on my mobile' campaign, by four European telecommunications firms – British Telecom, Nokia, Deutsche Telekom and Vodafone. The e-Sustainability Initiative's rather obtuse objective is to 'further sustainable development in the information and communication technology sector'.[2] One of its initiatives focuses on global supply chains. Material published by the e-Sustainability Initiative acknowledges that poor labour and environmental standards exist in the information and communications technology supply chain, and that there is 'mining in conflict zones for raw materials used in electronic equipment'. In 2010, the e-Sustainability Initiative had over twenty members, including Alcatel-Lucent, Huawei, Nokia, Cisco, Sony Ericsson, Hewlett Packard and Motorola. The Electronics Coalition is more broadly focused on the electronics industry compared to the e-Sustainability Initiative, which is focused on telecommunications. It was established in 2004 by three leading brand-name companies (Hewlett Packard, Dell and IBM) and five manufacturers of components for electronic products (Solectron, Sanmina-SCI, Jabil, Celestica and Flextronics). The Electronics Coalition developed a code of conduct for the electronics industry with sections on worker safety, fairness, environmental responsibility and business efficiency in the global electronics supply chain. The code addresses labour, health and safety, environmental, management systems and ethics issues, and member corporations' compliance with the code is audited. The Electronics Coalition helps members comply with the code by providing assistance and resources, as well as training in how the audit process works. In 2010 the Electronics Coalition had over forty members, including Cabot and Talison, and phone handset makers Sony Ericsson and Samsung.

In a development that increased the pressure on mineral trading corporations, in February 2007 Global Witness made a complaint to the British government's National Contact Point for the OECD's *Guidelines for Multinational Enterprises*. Global Witness alleged that Afrimex, a British-based minerals trading firm, had breached the guidelines by exporting 165 tonnes of coltan worth $2.5 million from the DRC between mid-1998 and November 2000 (Afrimex allegedly also exported cassiterite). During this period eastern DRC was controlled by forces from Rwanda, Uganda, RCD-Goma and RCD-ML, and Afrimex confirmed that it had paid taxes to the 'authorities', which by definition must have been armed groups deemed illegal by the United Nations.[3] Afrimex was a good target because it continued to export coltan from the DRC throughout the 2000s and was clearly indifferent to calls for companies not to trade in illegally exploited natural resources. The National Contact Point's final investigation report upheld the majority of the allegations and made recommendations to Afrimex for improving corporate social responsibility, including the development of a corporate social responsibility policy. Afrimex subsequently claimed that it no longer trades minerals from Central Africa, but this has not been independently verified and the National Contact Point is not set up to undertake compliance checks.

Other developments have also increased the pressure on corporations. The United Nations resolution 1856 from 2008, which expanded MONUC's mandate to include intervention against illegal armed groups' economic activities, raised the possibility of intervention against corporations trading coltan from armed groups. In early 2009 the Enough Project's conflict minerals campaign got under away and in July 2010 conflict minerals legislation in the United States was signed into law by President Obama. In the midst of these initiatives another prominent minerals trading firm, Traxys, pulled out of the DRC market, and the International Tin Research Institute (tin is one of the minerals targeted by the Enough Project and US legislation), supported by Electronics Coalition and e-Sustainability Initiative members, established a pilot tracing scheme for tin produced in the DRC.

The tantalum industry also decided to take action. The Tantalum-Niobium International Study Center adopted an artisanal and small-scale mining policy that states 'members will comply with host government laws and regulations', and which encourages artisanal and small-scale mining of tantalum in conditions of 'freedom, equality, safety and human dignity'.[4] The policy is aimed at encouraging members to avoid coltan that is produced, traded or exported by armed groups or associated companies named by the United Nations. However, the Tantalum-Niobium International Study Center has no power or resources to monitor adherence to the policy and is not set up to perform that role; compliance by members is voluntary. The Tantalum-Niobium International Study Center also established a working group on tantalum and niobium mining to consider members' options in regard to artisanal and small-scale mining and conflict minerals issues. Members, supported by the Electronics Coalition and e-Sustainability Initiative, agreed to fund an expansion of the pilot tracing scheme for tin to include coltan in 2010. If the tracing scheme works for coltan in the way it does for tin, it will commence by requiring exporters to complete paperwork attesting to the chain of custody for the minerals they export. A second phase will involve verifying the mines in which coltan is produced, followed by a third phase that verifies the provenance of minerals and implements a system of monitoring firms' performance against a range of standards including supply chain mapping, chain of custody, legitimacy, business ethics and avoiding any financing of armed groups. Like other industry initiatives, the pilot tracing scheme for tin does not include independent third-party verification and its implementation depends on continued funding by the industry.

The expansion of the tin scheme to include coltan raises questions about coordination with the German-funded certification project discussed in chapter 4. Both projects aim to do the same thing, although the tin initiative is focused on North Kivu and the German project is focused on South Kivu. Both have the support of the DRC government, and there is informal communication between the initiatives.[5] While it would be interesting to compare the results of an industry-run certification programme and a government-run scheme, there are risks associated with duplication.

Some members of the Tantalum-Niobium International Study Center could not be happier about the negative publicity surrounding tantalum sourced from Central Africa: these are mining companies with operations elsewhere. In one of the quirks of coltan politics, some mining executives have adopted activist arguments as part of their marketing pitch. The executive chairman of Noventa, which owns tantalum mines in Mozambique, stated: 'the repercussions for any company found using such disreputable material [i.e. coltan] in its high profile consumer electronics goods would be massive'.[6] Following the closure of Wodgina, the chief executive of Talison claimed: 'without Talison's supply the majority of

the world's tantalum will come from irregular and unreliable
suppliers from politically unstable regions, with much of it
coming from the DRC'.[7] It sounds as though even *he* believed
the DRC was home to 80 per cent of the world's tantalum
reserves! It is in the interests of companies such as Noventa
and Talison to feed corporate fears about consumer backlash,
United Nations sanctions and government action on coltan, if
they themselves produce tantalum from a competing source.

Scaremongering by Noventa and Talison illuminates one
consequence of natural resources activism: stable producers
of commodities targeted by activists do well out of these campaigns. First-World producers do particularly well because they
are better able to influence and document the conditions under
which a commodity is produced, as well as prove provenance.
For example, the Kimberley Process was a boon to producers
and marketers of diamonds from Australia, Canada and South
Africa – no 'blood diamonds' in these stable countries. In the
case of tantalum, producers elsewhere in Africa and in Canada,
Brazil and Australia stand to gain from initiatives to restrict
trade in coltan, at least once mines closed during the global
financial crisis reopen. It is no wonder that the coltan trader
from South Kivu quoted in chapter 4 viewed efforts to ban coltan
as an attempt by 'you people outside' to penalize Congolese.

The above developments are all in the sphere of international business. What changes have there been in the Congo
itself? Not many. MONUC, armed with its stronger mandate,
has made some limited seizures of illegally mined and traded
ore. In late 2008 and early 2009 exporters reportedly stopped
using the Bunagana and Ishasha border crossings as they
were aware that, if they used the crossings and paid any taxes
or export duties to the CNDP, they would be in violation of
international sanctions. This probably resulted in a diverting
of export routes rather than a cessation, but it demonstrates
that sanctions can have an effect on company practices. These

border crossings were presumably used once again follow-ing the demobilization of the CNDP and the absorption of its troops into the DRC army.

If there have been few changes to the way coltan is produced and traded in the Congo, yet many companies have said they have stopped buying or using tantalum from the DRC, where is all the coltan going?

Enter the dragon

The coltan is probably going to China. Chinese companies, backed by the government of the People's Republic of China, many of them state-owned corporations, have aggressively entered the global minerals and metals market. China has also become Africa's biggest trading partner. Two-way trade increased nine-fold from $10 billion in 2000 to $90 billion in 2009, surpassing US–Africa trade which was worth $86 billion in 2009.

Minerals in Africa have been a key target of Chinese mining corporations, exemplified by the $6 billion joint venture signed in October 2009 between the governments of China and the DRC. Under the terms of the deal, Chinese partners in the joint venture will build 4,000 kilometres of roads, 3,200 kilometres of railways, as well as 30 hospitals and 140 health centres worth about $3 billion in total. In return, the Chinese joint venture partners received the rights to build a new $3 bil-lion mine in Katanga and extract 6.8 million tonnes of copper and 430,000 tonnes of cobalt. The copper and cobalt is worth an estimated $80 billion at current prices. The new railway will link the copper mines of Katanga with the DRC's Atlantic port at Matadi, ending the country's dependence on foreign, especially South African, ports for mineral exports. The deal allows for a second phase of infrastructure projects worth an additional $3 billion if the first phase delivers as promised.

There could not be a bigger contrast to the efforts of H.C. Starck – one of the two behemoths of the tantalum industry (Cabot being the second) – to disengage with coltan, than the Chinese embrace of the DRC. The entry of Chinese firms into the African mining sector is partly the result of a 'go global' state policy driven by the Chinese government's efforts to secure supplies of raw materials to feed its booming economy. But it is also because Chinese corporations, especially those that are state-owned, are more willing to enter into deals with African governments than companies from North America, Europe and Japan, which are more likely to be publicly owned, risk-averse and mindful of reputational costs. Across Africa there is a vacuum of investment and technology, especially in countries whose governments have unpalatable records on corruption or human rights. The Chinese government is keen for Chinese firms to fill that void. Unlike the West's history of conquest and domination in Africa, China–Africa relations have an amicable foundation upon which to build. In the 1960s and 1970s the Chinese government supported Africans in their struggle against colonial rule and provided economic support for African governments (in return for political recognition of the communist government in Beijing). In the 1990s China's model of economic growth under autocratic rule, growth that moved millions of people out of poverty and resulted in extensive infrastructure, also appealed to many Africans and African rulers.

China itself produced about 8 per cent of the world's tantalum in 2008, but it is also a large importer and processor of tantalite – recall that Ningxia Orient Tantalum Industry is the third largest processor of tantalite after H.C. Starck and Cabot, and produces about 20 per cent of the world's tantalum powder. Direct imports of tantalum and other rare metal ores and concentrates from the DRC first show up in Chinese trade data in 2000, although imports from Rwanda appear to have commenced in 1996. In 2002, the United Nations found evidence

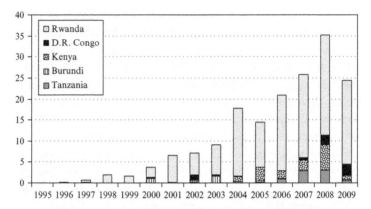

Source: Compiled by author from Trade Law Centre for Africa (2010), *TRALAC China Africa data.* Stellenbosch, South Africa, Retrieved 6 April 2010 from http://www.tralac.org/cgi-bin/giga.cgi?cat=1044&limit=10&page =0&sort=D&cause_id=1694&cmd=cause_dir_news#china

Figure 5.1 Imports to China of niobium, tantalum, vanadium and zirconium ore and concentrates, selected countries, 1995–2009, millions of dollars

that Eagle Wings, a Bukavu-based *comptoir*, sold coltan to Ningxia. The United Nations also identified a Ningxia report stating that '50 per cent of all coltan purchased for process-ing originates in Central Africa',[8] although Ningxia denied importing any coltan from the DRC. There are other reports of Chinese involvement: of dealings in 2004 between Chinese officials and Rwandan companies involved in the coltan trade; that almost all coltan from the DRC goes to China; and that Chinese processors turned to producers in Central Africa at the expense of Talison in Australia. By the end of the 2000s China had twenty plants capable of processing tantalum, but there were only ten plants elsewhere in the world. Figure 5.1 shows the value of imports of tantalum and three other minor metals (niobium, vanadium and zirconium) into China from countries whose ores are likely to originate in part or fully from the DRC.

It is not possible from these data to separate the value of imports of tantalum from that of niobium, vanadium and zirconium. However, given the size of the tantalum industries in the DRC and Rwanda, tantalum is likely to comprise the largest proportion of imports. Figure 5.1 shows a steady increase in the value of imports over the 2000s. In 2000 imports were worth $3.8 million, but nine years later they had increased eight-fold to $24.5 million. Assuming most of the value of imports is due to tantalum, the very high prices in 2000 compared to 2009 also suggest that the volume of tantalum being imported has increased dramatically. The decrease in 2009 may be due to a drop in the price of ores associated with the global financial crisis, rather than a decline in volume. While a minor proportion of ores and concentrates are imported directly from the DRC, most of those imported from Rwanda, Burundi, Tanzania and Kenya are likely to originate in the Congo. Despite reports of coltan being exported from Uganda during the Congo War, none of this appears to have been destined for China.

China's importance to the politics of coltan goes far beyond being an importer and processor of coltan. Manufacturing firms from China are expanding their market share further along the tantalum supply chain. Firms from Taiwan with which China has a symbiotic economic relationship, are already established giants in the telecommunications manufacturing world and produce about 50 per cent of all semiconductor chips and 90 per cent of all laptops, mostly in mainland China. Many other capacitor and circuitboard manufacturers are based in China. In 2009, eighty such suppliers received training in the Electronics Industry Citizenship Coalition's audit requirements. While some component manufacturers in China are subsidiaries of foreign firms, two wholly Chinese-owned firms, ZTE and Huawei, are amongst the world's top manufacturers of phone handsets. In 2009, ZTE was ranked no. 6 behind (in descending order) Nokia, Samsung, LG Electronics, Research in Motion and Sony Ericsson. In 2009, the South African multinational telecommunications firm, MTN, 'placed an order with a Chinese manufacturer to supply handsets at $13 each'[9] – a price African consumers love and with which Western brand-name companies cannot compete.

Future demand for tantalum

A key factor in whether coltan continues to be a source of profits for armed groups, and therefore the focus of activists, is future demand. Given that there is no looming long-term shortage of tantalum, notwithstanding temporary supply constraints associated with the closure of Wodgina and the exhausting of inventories, companies have few incentives to turn to alternatives to tantalum. For example, it is unlikely that ceramic or aluminium capacitors will replace tantalum capacitors, due to their higher cost. Without a shortage in the supply of tantalum there are also few incentives for further technological innovation to replace tantalum. In sum, the long-term prospects for tantalum appear good. Before the global financial crisis, demand for tantalum grew at 7 per cent to 10 per cent annually, and by 2012 there should be a full recovery in demand following the easing of the financial crisis.

Future demand will largely be dependent upon markets for electronic devices, given that about two-thirds of tantalum is used in tantalum capacitors, and these markets look set to expand. Mobile broadband (wireless connection to the Internet) alone is forecast to quadruple between 2009 and 2014 to almost 1.5 billion subscribers – that means an extra 1 billion laptops, mobile phones and personal digital assistants, a significant proportion of which will contain tantalum capacitors. Other potential growth markets include superalloys for the aeronautics industry, hard metal cutting tools used in the automotive industry and even hybrid vehicles. A new technology for hybrid vehicle batteries includes a thin layer of tantalum, which 'would result in an enormous cost advantage, and significantly reduce the amount of gold that is currently used for the same purpose'.[10] Ageing societies may also result in greater demand for implants containing tantalum, such as artificial hips.

Shifting consumer power

Activist campaigns have been effective in making mobile phones symbolic of both coltan and the connection between people around the world to violence and conflict in the Congo. In the author's own experience of explaining this project, the typical comment when people had heard of tantalum was 'Oh, you mean that mineral in my mobile phone'. However, as evidenced by calls to boycott mobile phones or to only buy telephones made using 'conflict-free minerals', mobile phones are a concrete element in tactical campaigns to pressure corporations by influencing the profits they make from selling telephones. The importance of mobile phones to these campaigns extends beyond their symbolic meaning, to activists' intention to reshape global telecommunications markets in order to facilitate peace in the DRC.

The rationale behind campaigns to get Western consumers to boycott mobile phones as a way of facilitating peace in the DRC is based on the power of these consumers to make corporations change their practices. The logic is this: if Western consumers refuse to buy mobile phones containing capacitors made with coltan, phone manufacturers will start making phones containing capacitors made from tantalum sourced from outside Central Africa. Corporations' refusal to buy coltan will, in turn, deprive armed groups of the revenue they use to wage war, thereby reducing violence.

One assumption underpinning this logic is that coltan is a sufficiently significant source of profits that, if deprived of it, armed groups will be forced to reconsider their violent tactics. Chapter 3 analysed this argument, and concluded that profits from coltan alone are not sufficiently important to change armed groups' calculations.

A second assumption is that Western consumers are powerful enough to change corporate behaviour because of the importance of rich country markets. Coltan campaigns imply that mobile phones are a plaything for wealthy consumers, highlighting even further, so the argument goes, the injustice of Congolese suffering harsh and violent conditions so people in rich countries can enjoy a gadget which brings few benefits to Congolese themselves. During her visit to eastern DRC in 2009, US Secretary of State, Hillary Clinton, stated: 'Every time someone uses a certain type of cell phone, they are using minerals that come right out of eastern Congo. What does that do for the people that I saw on the way from the airport into the city? Nothing. It helps them in no way.'[11] In fact, there is a good chance that many of the Congolese Ms Clinton saw on her way into town have their own mobile phone and use it to conduct business, keep track of children and stay in touch with relatives. As any visitor to the DRC quickly learns, mobile phones are everywhere. Market 'teledensity' (the number of mobile phones per 100 inhabitants) may not be as deep as in other African countries or rich countries, but mobile phones are ubiquitous.

Given that mobile phones have spread well beyond Western markets, how important are Western consumers to the mobile phone industry and therefore to manufacturers of electronic devices more generally? Will Western consumers' preferences continue to have the power to change corporate practices? Probably not. Due to Western nations' relative wealth, the first and largest markets for electronic devices such as mobile phones and computers have historically been in Europe, Japan and North America, but this is no longer the case. Consider the following:[12]

- in 2000, developing countries accounted for one-quarter of the world's 700 million mobile phones; by 2009 their share had grown to three-quarters (out of 4 billion);
- Africa has 35 per cent more mobile phone subscribers (364 million) than the United States (270 million);
- the DRC has more mobile phone subscribers (9.3 million) than either Switzerland (8.8 million) or Finland (6.8 million);
- there are twice as many Internet subscribers in China (150 million) than in the United States (72 million);
- Africa has more Internet subscribers (9.7 million) than Spain (9.3 million).

Table 5.1 shows the top twenty countries in terms of numbers of subscribers to mobile phones and Internet services. Subscriber data are inexact and indirect measures for numbers of mobile phone handsets and computers. A subscription to a mobile phone is actually to a subscriber identity module, or SIM, card and not the handset that contains the tantalum capacitor. This means that the number of handsets is less than the number of subscriptions in Table 5.1. Similarly, an Internet subscription might be used by more than one computer, e.g., a household might have one subscription but three computers, each of which contains tantalum. In this case, the number of computers is probably more than the number of Internet connections. Despite these qualifications, mobile phone and Internet subscriptions are useful proxies for the relative size of different national markets and therefore the power of consumers in different countries.

Rich countries are well-represented in both lists in Table 5.1, but so are developing countries – especially in the list for

Table 5.1   Top twenty countries (plus DRC) for mobile phone and Internet subscriptions, 2008[13]

Mobile phone subscribers (millions)		Internet subscribers (millions)	
1. China	634	1. China	150
2. India	347	2. United States	73
3. United States	271	3. Russia	31
4. Russia	188	4. Germany	20
5. Brazil	151	5. Britain	19
6. Indonesia	141	6. France	19
7. Japan	110	7. South Korea	15
8. Germany	107	8. India	13
9. Italy	89	9. Italy	12
10. Pakistan	88	10. Brazil	11
11. Thailand	79	11. Canada	10
12. Britain	76	12. Spain	9
13. Mexico	75	13. Mexico	8
14. Viet Nam	70	14. Australia	8
15. Philippines	68	15. Ukraine	6
16. Turkey	66	16. Taiwan	6
17. Nigeria	63	17. Netherlands	6
18. France	58	18. Turkey	6
19. Ukraine	56	19. Viet Nam	5
20. Spain	50	20. Malaysia	5
63. D.R. Congo	9	114. D.R. Congo	0.05

Source: International Telecommunications Union, Geneva (2009), ICT Statistics Database. Retrieved on 6 April 2010 from http://www.itu.int/ITU-D/ICTEYE/Indicators/Indicators.aspx#.

mobile phone subscriptions. Mobile technology has become popular in developing countries due to poor landline networks. The importance of developing country markets is partly a matter of demographics – sheer size of population – but also a matter of available infrastructure; for example,

Malaysia has a relatively a small population, 28 million in 2007, but a high rate of Internet penetration places it in the top twenty for Internet subscribers.

Due to higher incomes, rich consumers in wealthier countries will buy more technically sophisticated and more expensive telephones and computers compared to consumers in developing countries. Rich consumers are also more likely to replace their mobile phones and computers more frequently than consumers in developing countries. Both these factors will make the value of individual rich country markets more than that of most individual developing country markets. However, if mobile phones and computers are measures of where tantalum goes, and therefore which consumers have the power to change the tantalum industry, it is clear that developing countries are at least as important as rich countries.

In fact, when the potential growth of mobile phone and Internet markets is taken into account, consumers in developing countries are more important to the future of the tantalum industry than consumers in rich countries. Countries that have large populations and good prospects for future economic growth, but which currently have relatively low rates of teledensity, will be growth areas for mobile phones because more and more people will be able to afford these products. Table 5.2 compares the population, teledensity rates and forecast economic growth for 2010 of the ten most populous developing countries (plus DRC) with the ten most populous OECD member states (plus Russia).

Table 5.2 makes a simple point. By scanning data for population and teledensity it is apparent that, compared to OECD countries, there are a lot more people in developing countries compared with OECD countries who do not have a mobile phone; forecast economic growth rates suggest that many of these people will soon have the opportunity to buy one. The

Table 5.2 Estimated population, teledensity and forecast economic growth, selected countries

Developing countries	Population (millions), 2007	Teledensity per 100 inhabitants, 2008	Forecast economic growth, 2010–11	OECD member states (plus Russia)	Population (millions), 2007	Teledensity per 100 inhabitants, 2008	Forecast economic growth, 2010–11
China	1,329	47	10.0	USA	308	88	2.8
India	1,165	29	8.6	Russia	142	132	3.6
Indonesia	225	60	6.1	Japan	127	86	1.9
Brazil	190	78	4.8	Mexico	108	70	4.3
Pakistan	173	53	3.5	Germany	82	130	1.5
Bangladesh	158	28	5.6	Turkey	73	87	4.3
Nigeria	148	42	7.2	France	62	94	1.6
Philippines	89	76	3.8	Britain	61	124	1.9
Vietnam	86	79	6.3	Italy	59	150	1.0
Egypt	80	54	5.3	South Korea	48	94	4.8
D.R. Congo	63	14	6.2	Spain	44	111	0.3

Sources: Data on population: United Nations Development Programme, New York (2009), *Human Development Report*, Table L: population data. Retrieved 13 April 2010 from http://hdr.undp.org/en/reports/global/hdr2009/. Data on mobile phone subscriptions: International Telecommunications Union, Geneva (2009), ICT Statistics Database: Mobile cellular, subscriptions per 100 people. Retrieved on 6 April 2010 from http://www.itu.int/ITU-D/ICTEYE/Indicators/Indicators.aspx#. Data on forecast economic growth is from *World Economic Outlook: Rebalancing Growth* (International Monetary Fund, Washington DC, 2010). Retrieved 6 July 2010 from http://www.imf.org/external/pubs/ft/weo/2010/01/index.htm. Where teledensity per 100 inhabitants is more than 100, this means that the average inhabitant has more than one mobile phone in a year.

fact that some people in rich countries have more than one mobile phone (indicated by subscriptions exceeding 100 per inhabitant) and regularly change their mobile phones or upgrade to more advanced technology demonstrates that rich country markets will continue to grow, but more slowly than developing country markets.

It has been established that the market for tantalum is bright. However, as a result of demography, economic growth and unmet demand for telecommunications products, markets for products containing tantalum are shifting away from rich countries to developing countries, especially in Asia. These changes are important to coltan campaigns and other future campaigns focusing on natural resources, because consumer boycotts and protest is the flagship tactic of activists seeking to reshape natural resources supply chains. To be effective consumers in the markets that matter most to corporations need to participate.

Changing markets are important for another reason: if established corporations continue to dominate the market for tantalum base products and for manufactured goods containing these products, activists will continue to have leverage over them even if consumer markets shift. This is because many of the corporations are from the same countries as the activists, particularly the United States and Northern Europe (Finland, Sweden, Germany and the Netherlands). However, this is not the case for Japan where there has been almost no coltan-related activism, although in 2009 Osaka University participated in 'Congo Week' and the 'breaking the silence' campaign organized by Friends of the Congo. Corporations often care about their image in the eyes of their 'home' public, as this may have an impact on their share price. Negative publicity may also precipitate action by governments or regulatory organizations to investigate corporations' supply chains, generating yet more negative publicity. However, if the corporations that make products containing tantalum are also from developing countries, what impact will this have on consumer boycotts and other efforts by activists to reshape natural resources industries? Will these companies be susceptible to consumer and activist pressure?

Shifting markets, shifting activism?

China has a long history of activism and protest focused on domestic political, economic, social and environmental issues, including by students, farmers and workers. However, Chinese from the People's Republic do not have a history of engaging in transnational activism focused on issues beyond China. One reason for this is that the Chinese government constrains the activities of organizations that typically facilitate transnational activism in developing countries, such as churches, independent unions and international NGOs. Transnational networks with strong connections to China, such as Falun Gong, have been targeted by the Chinese government because it views these networks as a threat to state authority. Chinese processors, manufacturers and brand-name companies are consequently less subject to domestic activist pressure regarding their international activities, compared to corporations in Western democracies.

In contrast to China, other developing countries with large markets for products containing tantalum – such as Brazil, India, Pakistan and the Philippines – have highly organized and robust non-profit sectors. There is publicity and activism about a range of global justice and environmental issues, including deforestation, climate change, Third World debt and economic globalization. There has been strong support in the past from these countries for some transnational issues, including justice issues in Africa, such as anti-colonial struggles and the anti-apartheid movement. However, activism in developing countries is often focused on the effects of global issues on the home front, such as conditions for workers in the factories of multinational corporations. One likely reason for the lack of engagement with transnational natural resource issues is that low incomes tend to make people less concerned about the origin and provenance of the products they buy.

These are the twin challenges faced by future campaigns to reshape the supply chains for coltan and other natural resources: consumers in developing countries uninterested in and uneducated about a war in distant Congo, and manufacturers in China of tantalum base products and electronics containing tantalum indifferent to pressure about their buying practices. Activists interested in coltan will probably continue to focus on Western consumers because they are accessible and there is a network of like-minded organizations through which publicity material can be circulated. However, as consumers in developing countries become more and more important to manufacturers, activist influence over these firms will decline unless activists extend their campaigns to developing countries.

Conclusion

This volume began by asking why an obscure mineral unknown to the general public a decade ago has become so contested. The issues that define coltan politics – armed groups' efforts to control and tax production and trade in the DRC, and international contestation between activists and corporations over the global supply chain – are a story of globalization. The locking up of tantalite production elsewhere in the world into long-term contracts in the late 1990s created perceptions of a shortage of supply, which caused the price boom in 2000. Coltan was one of the few sources of tantalite able to fulfil spot market demand. Armed groups waging war in the DRC had an interest in exploiting any and all sources of income, and the sudden profitability of coltan temporarily made it highly attractive. The extent to which armed groups have been able to profit from coltan is directly related to the political and economic conditions in which they operated: a weak state unable to protect property or enforce contracts, an economy based on cheap labour, and a mineral amenable to artisanal production.

Contestation between armed groups occurs over other minerals found in eastern DRC, and in some other countries, so why have gold, tin or tungsten not achieved the infamy of coltan, despite being equally attractive to armed groups? Why has coltan developed the politics it has? First, the price boom caused a 'coltan rush' in the Congo, so there was a very real movement of people and armed groups to coltan deposits and the evolution of a domestic politics of coltan focused on contestation over the control of production and trade. Second, the very obscurity of tantalum made a good tale: of the Congo, in darkest Africa, having the mother lode – the 80 per cent claim – of a rare resource desperately needed by Westerners to sustain their indulgent lives, but fought over by savage tribes. A tired refrain, but one that sells newspapers and appeals to readers. Third, the decision by a Belgian activist in 2001 to make mobile phones the symbol of the connection between consumers and war was a fateful one for the tantalum industry, activists interested in the DRC and brand-name corporations.

A politics of coltan has not arisen just because it makes a good story, although the drama of coltan has been instrumental in the development of an international dimension to the mineral's politics. Coltan *has* shaped the interests and strategies of armed groups in eastern DRC, and these groups have profited from it. Yet, following the price boom, coltan returned to being just one of many sources of income for armed groups, including gold, tin, tungsten, manganese, timber, cattle, other livestock, wildlife, taxation of commerce, theft of consumer goods, theft of agricultural produce and control of international border posts. Coltan continues to be a specific factor in the calculations of many individual miners, *négociants* and *comptoirs* and continues to earn profits for some armed groups. However, it will be more and more difficult to distinguish a domestic politics of coltan from the broader resource politics of eastern DRC and Central Africa unless there is another price boom that results in renewed contestation that is distinct from contestation over other resources. In the future, scholars may well view the politics of coltan as a minor chapter in a complicated war that was never just about resources.

The potential demise of a domestic politics of coltan is irrelevant to coltan's other life: as an international cause célèbre, through its association with mobile phones, to injustice and violence. Campaigns focused on coltan have made a difference to the global tantalum supply chain in that Western firms appear to be avoiding coltan in favour of tantalite from other sources. But this has resulted in a diverting of supply routes to China, rather than a cessation of trade from the DRC. The pressure on Western corporations to avoid coltan will intensify as the result of legislation in the United States and the ongoing efforts of campaigns such as those by the Enough Project and Friends of the Congo. This pressure will create further opportunities for Chinese processing and manufacturing firms to increase their role in the global tantalum industry, in keeping with the Chinese government's goal of dominating the global minor metals and rare earth markets.

The unknown factor in Chinese involvement in the trade in coltan and other conflict minerals is whether Beijing's quest for 'soft power' – influence in global affairs achieved through non-military means, such as culture, political values, foreign policy and economic activity – will result in the government pressuring Chinese firms to desist from certain activities in order to shore up the image of China as a responsible and law-abiding power. Official Chinese economic engagement with Sudan and Burma, notwithstanding the international community's pillorying of those two countries' regimes and concerns about ongoing violence at the hands of government or government-allied armed forces, suggests the Chinese government will pay little attention to claims that Chinese firms buying or processing coltan are engaging in illegal trade or fuelling conflict.

Initiatives focused on the production and domestic trade of coltan, such as certification and fingerprinting schemes and MONUC's seizure of illegal consignments, have the potential to formalize and increase transparency in the trade, making it more difficult to sell ore that has passed through the hands of armed groups. However, having been burnt by accusations that they are fuelling conflict, major processing firms may need incentives, such as cheaper prices, to buy even certified coltan. Consumers might accept the greater expense of fair trade products, but given that tantalum goes through so many stages of processing and then becomes part of a complex product, a market for fair trade coltan is unlikely to develop. Besides, all the other materials in a mobile phone, for example, would also have to be fair trade.

But let us get to the heart of the matter. *What can and should a concerned person do to end the relationship between coltan and war in eastern DRC?* The most important step is to learn the facts about natural resources and conflict, so any actions and decisions are made from an informed position. Read, starting with the selected readings at the end of this volume. Specifically, is it worth participating in a mobile phone boycott? If the aim is to bring attention to the war in the Congo, yes, definitely; but if the aim is to end the war, no. Consumer boycotts will not end the war, because the war is not just about resources. If coltan – or even all minerals – were taken out of the equation violence would continue, motivated by decades-old grievances around land and citizenship. *Should coltan be banned?* If the purpose of a ban is to make manufacturing firms pay attention to where minerals in their supply chain come from, without any specific goal of changing conditions on the ground, yes; if the purpose is to end the war in the Congo, no. Reducing the income of armed groups is a good thing, but it may not be possible to achieve, given that corruption and porous borders will mean that illegal coltan will always find a buyer, probably from China. Armed groups and their individual members need incentives to leave the minerals trade, and this probably means provision of alternative economic opportunities. Successful interdiction of illegally exported coltan will, at the very least, require cooperation from governments in the region, especially the Rwandan government. However, the last thing ordinary Congolese need is a state that is more interventionist in domestic and international commerce. *What are the means to bring about the ending of the war?* In addition to needing an army that is paid by the state instead of engaging in economic activities to generate income (including coltan mining), the DRC needs an army with a functioning hierarchy and good lines of command and control that is capable of militarily defeating anti-government forces, and a justice system capable of enforcing property rights and contracts. Getting armed groups to stop fighting may involve pyrrhic compromises such as amnesties and incentives for individual combatants to demobilize. Further, the government needs to address the grievances around land and citizenship that are key causes of violence in eastern DRC.

There are three lessons from the politics of coltan for the politics of natural resources more generally. First, companies should be aware that, no matter how innocuous they think their business is and how mundane or popular their products are, with time and money activists can create and propagate a compelling narrative of injustice that implicates their firm – which becomes a simple task when the firm actually buys commodities produced under conditions of conflict. Second, while non-state actors such as NGOs and multilateral organizations have successfully carved out a role in natural resource politics, this role can never be assured. The increasing importance of Chinese firms, backed by the Chinese state, in the global minerals and metals markets demonstrates that corporations and states will remain formidable actors. Third, the shifting markets for tantalum demonstrate that the companies and consumers of Western countries are less important to the politics of natural resources than they used to be, and their importance will further diminish. If Western activists want to maintain the influence they have fought so long and hard to obtain, they need to find messages that resonate with consumers in developing countries, work out methods of putting pressure on corporations whose domestic reputation is little affected by their actions overseas, and seek ways of persuading governments, which may have a tradition of not engaging with domestic political issues in other countries, to care about what their corporate citizens are doing.

Having discussed the array of challenges facing future advocacy about natural resource politics, it is worth returning to the energy and determination of activists. It is these individuals and organizations that took up the issue of conflict minerals exposed by the United Nations a decade ago, and in terms of raising awareness of the conflict in the Congo they have succeeded. *They did it!* There are inaccuracies and exaggerations in their stories of cause and effect about coltan and conflict, but whether these narratives are accurate or not, and whether corporations like it or not, activists and NGOs are determined to politicize the exploitation of natural resources such as coltan because of the relationship they perceive to inequity and violence in developing countries. And they are driven by two essential qualities: patience and optimism:

> Forty years ago the situation for diamonds was like that for coltan today . . . Protest and awareness will change the way coltan is mined and sold, and the way the money from it is passed on. It's not going to happen tomorrow, but I feel optimistic. Some of us can participate to speed up the process.[14]

Activists of the past harnessed these qualities, and proved capable of strategically adapting and learning new tactics when needed. The activists of the future will need to do the same, for tomorrow is going to be a new world. The basic political elements of natural resource politics will remain the same – business, consumers, governments, civil society, supply and demand – but the geography of power is set to radically change.

Notes

CHAPTER I

1 Senator Richard Durbin (Democrat, IL), United States Senate Majority Whip, 24 April 2009.

2 Confusion over names extends to 'columbite'. Columbium is the old-fashioned name for the metal derived from columbite, which has the atomic number 41. In 1950 the International Union of Pure and Applied Chemistry officially designated this element 'niobium', instead of 'columbium', and gave it the symbol Nb.

3 Prices are an average of international spot market prices. They are adjusted for inflation using deflators from the Development Co-operation Directorate of the OECD, Paris. Retrieved on 2 February 2010 from http://www.oecd.org/document/6/0,3343, en_2649_34447_41007110_1_1_1_1,00.html.

4 The discussion on the causes and consequences of the 1978–80 price spike is based on the author's correspondence with Emma Wickens (Secretary General, Tantalum-Niobium International Study Center) on 6 April 2010 and 24 March 2010.

5 Author conversation with Isaac Djumapili, a former coltan trader from Bunyakiri, South Kivu. Sydney, 20 February 2010.

6 This information is based on author correspondence with Emma Wickens (Secretary General, Tantalum-Niobium International Study Center), 24 March 2010. The media articles referred to are: Astill, James & McKie, Robin (2001), Gorillas face doom at gunpoint. *Guardian*, 4 March 2001. Retrieved 13 March 2010 from http://www.guardian.co.uk/world/2001/mar/04/ robinmckie.jamesastill; Bond, Michael and Braeckman, Colette (2001), A moral minefield. *New Scientist* **2285**, 7 April; and Vesperini, Helen (2001), Congo's coltan rush. *BBC News*.

Retrieved 1 August 2009 from http://news.bbc.co.uk/2/hi/
   africa/1468772.stm.
7 Garrett, Nicolas and Mitchell, Harrison (2009), *Trading Conflict
   for Development: Utilising the trade in minerals from eastern DR
   Congo for development.* Resource Consulting Service, Aston
   Sandford, Great Britain, and the Communities and Small-Scale
   Mining Initiative, Washington DC.
8 Reuters (2009), Australia's Talison to restart Wodgina
   tantalum mine. Online article, 23 September 2009.
   Retrieved 6 April 2010 from http://www.reuters.com/article/
   idUSSYD47982820090923.
9 Metals Place (2008), A steep price increase is expected for
   tantalum. Retrieved 18 July 2009 from http://metalsplace.com/
   news/articles/21449/a-steep-price-increase-is-expected-for-
   tantalum/; and Vulcan, Tom (2009), Tantalum: A Modern Metal,
   Actually. 13 January 2009. Retrieved 6 April 2010 from http://
   www.hardassetsinvestor.com/features-and-interviews/1/1376-
   tantalum-a-modern-metal-actually.html.

CHAPTER 2

1 World Bank, Washington DC (2008), *Democratic Republic of
   Congo: Growth with governance in the mining sector.* Report no.
   43402-ZR, p. 56.
2 Interview with Halera (16) and Safari (17), former schoolchildren
   now mining coltan. Interview 2.6 in Tegera, Aloys, Mikolo, Sofia
   and Johnson, Dominic (2002), *The Coltan Phenomenon: How a
   rare mineral has changed the life of the population of war-torn North
   Kivu province in the east of the Democratic Republic of Congo.* Pole
   Institute, Goma, D.R. Congo, p. 15.
3 Interview 2.4 with Alphonse Batibwira, a teacher in Matanda
   village. Ibid., p. 15.
4 Author conversation with Isaac Djumapili, a former coltan trader
   from Bunyakiri, South Kivu. Sydney, 20 February 2010. For other
   information about measuring coltan at mine sites see Redmond,
   Ian (2001), *Coltan Boom, Gorilla Bust. The impact of coltan mining
   on gorillas and other wildlife in eastern DR Congo.* Dian Fossey
   Gorilla Fund, Atlanta, & Born Free Foundation, Horsham, UK,
   p. 9; and Tegera et al. (2001), The Coltan Phenomenon, p. 19.

5 Tegera et al. (2001), *The Coltan Phenomenon*, p. 19.
6 See Redmond (2001), *Coltan Boom, Gorilla Bust.* pp. 9–10.
7 Interview 1.4 with Bernard Luanda (President, Bushenge/Hunde mutual aid group). Tegera et al. (2001), *The Coltan Phenomenon*, p. 12.
8 Redmond (2001), *Coltan Boom, Gorilla Bust.* p. 9.
9 Author correspondence with Harrison Mitchell, Resource Consulting Services, 22 April 2010.
10 Author correspondence with coltan trader, 22 February 2010.
11 Gippsland Limited, Perth (2007), Ten Year Tantalum Offtake Contract. Press release, 13 November.
12 Steinweg, Tim and de Haan, Esther (2007), Capacitating electronics: the corrosive effects of platinum and palladium mining on labour rights and communities. makeITfair, Amsterdam, p. 9.

CHAPTER 3

1 Michael Nest (2006), The political economy of the Congo War. In: Nest, Michael, with Grignon, François and Kisangani, Emizet F., *The Democratic Republic of Congo: Economic dimensions of war and peace.* Lynne Rienner, Boulder CO, p. 43.
2 Collier, Paul (2000), Doing well out of war: an economic perspective. In: Berdal, Mats and Malone, David M. (eds.) *Greed and Grievance: Economic Agendas in Civil Wars.* Lynne Rienner, Boulder, Colorado, p. 94.
3 International Committee of the Red Cross, Geneva (2002), *War, Money and Survival*, p. 45.
4 See Human Security Report Project (2010), ch. 3: The Death Toll in the Democratic Republic of the Congo. In: 'Shrinking Costs of War', Part II in the Human Security Report 2009. HSRP, Vancouver, pp. 36–48; and International Rescue Committee and Burnet Institute (2008), *Mortality in the Democratic Republic of Congo: An ongoing crisis.* New York. Estimates include both direct and indirect deaths.
5 Polgreen, Lydia (2008), Congo's riches, looted by renegade troops. *New York Times*, 16 November 2008. Retrieved 6 April from http://www.nytimes.com/2008/11/16/world/africa/16congo.html, p. 5. For more information on taxes levied

by the RCD-Goma see de Failly, Didier (2001), Coltan: pour comprendre. In: *L'annuaire des Grands Lacs*, L'Harmattan, Paris, pp. 279–306.

6 Author correspondence with Emma Wickens (Secretary General, Tantalum-Niobium International Study Center), 24 March 2010.

7 Werner, Klaus and Weiss, Hans (2002), *Schwarzbuch markenfirmen. Die machenschaften der weltkonzerne* (The Black Book on Brand Companies). Franz Deuticke Verlagsgesellschaft, Vienna.

8 See the International Peace Information Service's *Interactive map of militarised mining areas in the Kivus* (Ghent, Belgium). Retrieved 6 April 2010 from http://www.ipisresearch.be/mining-sites-kivus.php.

9 Vick, Karl (2001), Vital Ore Funds Congo's War. *Washington Post*, 19 March 2001, p. A01.

10 Garrett & Mitchell (2009), *Trading Conflict for Development*, p. 6.

11 Estimate of $62.6 million for the Rwandan army for 1999 is based on UN data (Report S/2001/357, p. 28). The UN estimated that over 18 months from January 1999 to June 2000, coltan sold by the Rwandan army was worth $250 million based on coltan prices of $200 per kilogram. The author calculated an average monthly figure and multiplied this by twelve to get an approximate annual figure for 1999 ($167 million). The author considers the price of $200 per kilogram to be an overestimate for 1999, and used estimates from the US Geological Survey of average spot market prices for 1999 ($74.95 per kilogram) to revise downwards the $167 million figure to $62.6 million. The estimate of $11.8 million for all armed groups for 2008 is the mid-range figure from Enough Project, with the Grassroots Reconciliation Group, Washington DC (2009), *Comprehensive approach to Congo's conflict minerals*.

12 Original data for 2005 were for January to June 2005, from Hayes, Karen, Hickock Smith, Kimberly, Richards, Simon and Culp Robinson, Richard (2007), *Researching Natural Resources and Trade Flows in the Great Lakes Region*. PACT, Washington DC. The author has doubled these to obtain a whole year estimate for 2005. Original data for 2008 were inflated by 35% by the Enough Project, based on estimated under-declaration rates for exports (see A Comprehensive Approach to Congo's Conflict Minerals, Enough Project with the Grassroots Reconciliation Group,

Washington DC, Endnote 1, p. 18). The author has revised the
Enough Project's data for value of exports for 2008 downwards to
take out the 35 per cent. Therefore, data for both 2005 and 2008
are underestimates and have not been adjusted for an estimated
export under-declaration rate.

13 Author correspondence with Dr Laura Seay, Morehouse College,
Atlanta, 5 April 2010.

14 BBC Television, *Heart of Darkness*. Broadcast on Australian
Broadcasting Corporation Television's 'Four Corners program'.
Retrieved on 30 July 2010 from http://www.abc.net.au/iview/#/
view/607757.

15 Ibid.

CHAPTER 4

1 Pisa, Nick (2009), Catholic bishops urge Italians to make the
ultimate sacrifice for Lent: give up texts, iPods and Facebook.
Daily Mail Online, 4 March 2009. Retrieved 6 April 2010 from
http://www.dailymail.co.uk/news/worldnews/article-1158928/
Bishops-urge-Catholics-make-ultimate-sacrifice-Lent-Give-texts-
iPods-Facebook.html.

2 For clarity and brevity United Nations reports are referred to
by their month, year and number, rather than their full title.
Resolutions are referred to by their number and year.

3 Enough Project, Washington DC (2010), *Raise Hope for
Congo* website, retrieved 26 March 2010 from http://www.
raisehopeforcongo.org/special-page/take-action-congo.

4 GSM or Global System for Mobile Communications was the term
used to describe the second generation of mobile telephones in
use in the early 2000s.

5 Friends of the Congo, Washington DC (2010), Retrieved 26
March 2010 from http://www.friendsofthecongo.org/about/
index.php.

6 Zoos Victoria, Melbourne (2010), Retrieved 11 March 2010 from
www.zoo.org.au/Calling_on_You.

7 Meyer, David S. and Corrigall-Brown, Catherine (2005),
Coalitions and political context: US movements against wars in
Iraq. *Mobilization* **10**(3), 327–44.

8 Campbell, Eric (2009), *Foreign Correspondent: The Congo*

Connection. Australian Broadcasting Corporation Television, 8 September 2009. www.abc.net.au.

9 Representative of a *comptoir*, Goma 9 August 2008, quoted in Global Witness (2009), *Faced with a gun what can you do? War and the militarization of mining in Eastern Congo*, p. 42.

10 President, Congolese Community of Australia Inc., who participated in a focus group on the motivations and logic of activists involved in coltan-related advocacy, held 6 February 2010 in Sydney.

11 Author conversation with Isaac Djumapili, a former coltan trader from Bunyakiri, South Kivu. Sydney, 20 February 2010.

12 Community organizer, refugee rights and Congolese issues, who was a participant in a focus group on the motivations and logic of activists involved in coltan-related advocacy, held 6 February 2010 in Sydney.

13 Activist A, member of a Christian church network and participant in a focus group on the motivations and logic of activists involved in coltan-related advocacy, held 6 February 2010 in Sydney.

14 Community organizer, refugee rights and Congolese issues, who was a participant in a focus group on the motivations and logic of activists involved in coltan-related advocacy, held 6 February 2010 in Sydney.

15 Retrieved on 31 July 2010 from http://www.enoughproject.org/blogs/congress-tackles-conflict-minerals-financial-reform-bill.

CHAPTER 5

1 Interviews with 'industry stakeholders' at the World Bank's Extractive Industries Week, March 2009. In Garrett, Nicolas and Mitchell, Harrison (2009), *Trading Conflict for Development: Utilising the trade in minerals from eastern DR Congo for development*. Resource Consulting Service, Aston Sandford, UK, and the Communities and Small-Scale Mining initiative, Washington, DC, p. 12.

2 www.gesi.org, accessed 2 April 2010.

3 Global Witness, London (2007), *Complaint to the UK National Contact Point under the Specific Instance Procedure of the OECD Guidelines for Multinational Enterprises.*

4 Tantalum-Niobium International Study Center, Brussels (2009), *Artisanal and Small Scale Mining Policy*, www.tanb.org, accessed 2 April 2010.
5 Author communication with Dirk Küster, Federal Institute for Geosciences and Natural Resources (BGR), 8 April 2010.
6 Clinton Wood (Executive Chairman, Noventa) quoted in Ruffini, Antonio (2008), Africa remains key to future tantalum supply. *Engineering and Mining Journal*, September, p. 70.
7 Talison Incorporated (2008), *Talison to suspend Wodgina tantalum operations*. Press release, 26 November 2008. Retrieved on 6 April 2010 from http://www.talisontantalum.com/pdfs/Media_Release_26_Nov_08_FINAL.pdf.
8 United Nations, New York (2002), *Final report of the panel of experts on the illegal exploitation of natural resources and other forms of wealth of the D.R. Congo*. Report S/2002/1146, pp. 16–17.
9 *The Economist*, (2009), Mobile marvels: A special report on telecoms in emerging markets. 26 September 2009, p. 6.
10 Firman, Carl (2008), Tantalum: Charged for business. *Global Capital Magazine* 23–6 October 2008, p. 24.
11 Hillary Clinton, US Secretary of State, during interview with Radio Okapi, Kinshasa, 10 August 2009.
12 Data are for 2008. International Telecommunications Union, Geneva (2009), *ICT Statistics Database*. Retrieved on 6 April 2010 from http://www.itu.int/ITU-D/ICTEYE/Indicators/Indicators.aspx#.
13 Data for Internet subscribers in Japan are unavailable for 2008. Based on the number of Internet users, Japan is likely to have the third largest number of subscribers.
14 Activist B, member of a Christian church network and participant in a focus group on the motivations and logic of activists involved in coltan-related advocacy, held 6 February 2010, Sydney.

Selected readings

Good data on tantalum are scarce and most data are incomplete. The most reliable data on global production are those produced by the Tantalum-Niobium International Study Center (www.tanb.org) and from Roskill Information Services' annual publication, *The Economics of Tantalum* (London, Roskill). For analysis of causes of historical price trends, see Larry D. Cunningham, Tantalum. In: *Metals Prices 1998* (Washington DC, United States Geological Survey (USGS), 1998); John J. MacKetta, *Supercritical Fluid Technology, Theory and Application to Technology Forecasting* (New York, Marcel Dekker, 1996). The minerals yearbooks and country reports of the USGS (www.usgs.gov) contain summaries of events affecting supply and demand, as well as data on tantalum production, reserves, trade and prices by country. However, USGS data are often incomplete, often omitting statistics for production and reserves in Africa, Russia and China. Data are also sometimes presented at the annual Minor Metals and Rare Earths Conference. For a excellent discussion of how mineral reserves are estimated see Richard Burt, 'Tantalum – a rare metal in abundance?' *TIC Bulletin* (2000) **141**, pp. 2–7.

Chapter 2 on the organization and production of the tantalum and coltan trade drew on two research projects undertaken during the tantalum price boom of 2000 and 2001, both of which involved the collection of qualitative data through field interviews. See Ian Redmond, *Coltan Boom, Gorilla Bust. The impact of coltan mining on gorillas and other wildlife in Eastern DR Congo* (Atlanta, Dian Fossey Gorilla Fund, and Horsham, Great Britain, Born Free Foundation, 2001), and Aloys Tegera, Sofia Mikolo and Dominic Johnson, *The Coltan Phenomenon: How a rare mineral has changed the life of the population of war-torn North Kivu Province in the East of the Democratic Republic of Congo* (Goma, Pole Institute, 2002). The following reports are also useful: Didier de Failly, Coltan: pour comprendre. In: *L'Annuaire des Grands Lacs* (Paris, L'Harmattan, 2001), pp. 279–306; Kevin D'Souza, *Scoping Study on the Artisanal Mining of Coltan in the Kahuzi Biéga National Park* (Atlanta, Dian Fossey Gorilla Fund, 2003); Karen Hayes and Richard Burge, *Coltan Mining in the Democratic Republic of Congo: How tantalum-using industries can commit to the reconstruction of the DRC* (Cambridge, Fauna & Flora International, 2003); Amis de la Forêt et de l'Environnement pour le Développement, *Rapport général de l'étude sur l'exploitation artisanale des ressources minières en territoire de Walikale: cas du coltan* (Goma, 2007); Philippe Le Billon and Christian Hocquard, Filières industrielles et conflits armés: le cas du tantale dans la region des Grands Lacs, *Écologie & Politique* 34, 2007, 83–92; Pole Institute, Blood Minerals: The criminalization of the mining industry in eastern DRC (Goma, 2010). The following media reports describe aspects of life in coltan mining communities: Blaine Harden, The Dirt in the New Machine. *New York Times Magazine*, 8 December 2001, 34–9; Eric Campbell, *Foreign Correspondent: The Congo Connection*. Australian Broadcasting Corporation Television (www.abc.net.au), broadcast 8 September 2009. For specific information on the impact of mining and war on national parks, see Dirk Draulans and Ellen Van Krunkelsven, The impact of war on forest areas in the Democratic Republic of Congo, *Oryx* 36, 2001, 37; Redmond, *Coltan Boom, Gorilla Bust* (2001); International Union for the Conservation of Nature, *Coltan Mining in World Heritage Sites*

in the Democratic Republic of Congo (Gland, Switzerland, 2001)
p. 1; D'Souza, *Scoping Study on the Artisanal Mining of Coltan
in the Kahuzi Biéga National Park* (2003). The Communities
and Small-Scale Mining network website (www.artisanalmin-
ing.org) has links to information and articles about artisanal
and small-scale mining, including locations, population data,
the minerals extracted using these methods, as well as political, economic and environmental analyses of such methods.

For an introduction to the importance of property rights and
their role in economic development, see Hernando De Soto,
*The Mystery of Capital: Why capitalism triumphs in the West and
fails everywhere else* (New York, Basic Books, 2000), especially
chapter 3. For an excellent, but academic, introduction to the
role of the state in protecting property rights, see section IV of
Thrainn Eggertsson's *Economic Behavior and Institutions* (New
York, Cambridge University Press, 1990). For a discussion of
regime type and property rights, see chapters 3 and 4 of Mancur
Olson, *Power and Prosperity: Outgrowing communist and capital-
ist dictatorships* (New York, Basic Books, 2000). For a discussion
on the role of ethnic networks in enforcing property rights
and contracts, see Avner Greif, Contract Enforceability and
Economic Institutions in Early Trade: The Maghribi Traders'
Coalition, *American Economic Review* 83(3), 1993, 525–48.

Chapter 3 brings together scholarship on war, natural
resources and the Congo. For an introduction to the relationships between conflict, natural resources and poverty, see Paul
Collier's *The Bottom Billion: Why the poorest countries are failing
and what can be done about it* (Oxford, Oxford University Press,
2007), and Patrick Regan's *Sixteen Million One* (Boulder CO,
Paradigm, 2009); Mats Berdal and David M. Malone (eds.),
Greed and Grievance: Economic agendas in civil wars (Boulder
CO, Lynne Rienner, 2009); Michael L. Ross, What do we
know about natural resources and civil war? *Journal of Peace
Research* 41(3), 2004, 337–56.

I argue that since 1989 eastern DRC has been affected by five waves of violence. The following selected readings provide an introductory analysis to the different waves of violence. For the first wave (ethnic-based violence in the early 1990s), see Jean-Claude Willame (ed.) *Zaire: Predicament and prospects* (New York, Minority Rights Group, 1997); Mahmood Mamdani, *When Victims Become Killers: Colonialism, nativism, and the genocide in Rwanda* (Princeton, Princeton University Press, 2001). For the second wave (spillover effects of the Rwanda genocide), see Timothy Longman, *Zaire: Forced to Flee. Violence Against the Tutsis in Zaire* (New York, Human Rights Watch, 1996); Jean-Claude Willame, *Banyarwanda et Banyamulenge: Violences ethniques et gestion de l'identitaire au Kivu* (Tervuren, Institut Africain-CDEAF, 1997); Mamdani, *When Victims Become Killers* (2001); Gérard Prunier, *Africa's World War: Congo, the Rwandan genocide, and the making of a continental catastrophe* (New York, Oxford University Press, 2008). For the third wave (violence associated with the war against Mobutu), see William Reno, Sovereignty and Personal Rule in Zaire, *African Studies Quarterly* 1(3), 1997; Gautier de Villers and Jean-Claude Willame, in collaboration with Jean Omasombo Tshonda and Eric Kennes, *Chronique politique d'un entre-deux-guerres: Octobre 1996–Juillet 1998*. Cahiers Africains 35–6 (Tervuren, Institut Africain-CEDAF, 1998); Mel McNulty, The Collapse of Zaire: Implosion, Revolution or External Sabotage? *Journal of Modern African Studies* 27(1), 1999, 53–82. For the fourth wave (the Congo War), see John F. Clark (ed.), *The African Stakes of the Congo War* (New York, Palgrave Macmillan, 2002); Michael Nest, with François Grignon and Emizet F. Kisangani, *The Democratic Republic of Congo: Economic dimensions of war and peace* (Boulder CO, Lynne Rienner, 2006); Thomas Turner, *The Congo Wars: Conflict, myth and reality* (New York, Zed Books, 2007). For the fifth wave (local-level violence), Séverine Autesserre's *The*

Trouble with the Congo: Local violence and the failure of international peacebuilding (London, Cambridge University Press, 2009) contains an excellent readable analysis of the extent and the causes of local violence in eastern DRC and the reasons local violence has been ignored by the international peacebuilding community.

For analysis of the causes of violence against civilians during civil war in general, see Stathis N. Kalyvas, *The Logic of Violence in Civil War* (New York, Cambridge University Press, 2006); Kristine Eck and Lisa Hultman, One-Sided Violence, *Journal of Peace Research* **44**(2), 2007, 233–46; Amnesty International, *Democratic Republic of Congo: North-Kivu: Civilians pay the price for political and military rivalry* (New York, 2005). For an analysis of the causes of sexual violence in general, see Elisabeth Jean Wood, Variation in Sexual Violence during War, *Politics & Society* **43**(3), 2006, 207–341. For analysis of the impact and extent of sexual violence in the DRC, see Human Rights Watch, *The War within the War: Sexual violence against women and girls in Eastern Congo* (New York, 2002); Marie Claire Omanyondo Ohambe, J.B.B. Muhigwa and B.M.W. Mamba, *Women's Bodies as a Battleground: Sexual violence against women and girls during the war in DRC, South Kivu (1996–2003)* (London, Réseau des Femmes pour un Développement Associatif, 2004); Maria Ericksson Baaz and Maria Stern, Why Do Soldiers Rape? Masculinity, Violence and Sexuality in the Armed Forces in the Congo, *International Studies Quarterly* **53**, 2009, 495–518. Also see BBC Television's *Heart of Darkness*, broadcast on Australian Broadcasting Corporation Television's *Four Corners* programme on 26 July 2010. For information on mortality caused by conflict in the DRC, see International Rescue Committee & Burnet Institute, *Mortality in the Democratic Republic of Congo: An ongoing crisis* (New York, 2008), and Human Security Report Project, The Death Toll in the Democratic Republic of Congo. Ch. 3

in 'Shrinking Costs of War', Part I in the Human Security Report 2009 (Vancouver, 2010), 36–48.

For information about which armed groups control coltan and other mines and the location of these mines, see the *Interactive map of militarised mining areas in the Kivus* at www.ipisresaerch.be, which is produced and maintained by the International Peace Information Service (Ghent, Belgium). For a discussion of the profits received by armed groups from coltan and other minerals, see Enough Project, *Comprehensive approach to Congo's conflict minerals – strategy paper* (Washington DC, 2009), available online from www. enoughproject.org.

For extended analysis of corporate involvement in the coltan trade during the Congo War period, see the following United Nations reports: *Report of the panel of experts on the illegal exploitation of natural resources and other forms of wealth of the D.R. Congo* (S/2001/357); *Addendum to the report of the panel of experts on the illegal exploitation of natural resources and other forms of wealth of the D.R. Congo* (S/2001/1072); *Final report of the panel of experts on the illegal exploitation of natural resources and other forms of wealth of the D.R. Congo* (S/2002/1146); *Final report of the group of experts on the Democratic Republic of Congo* (S/2008/773). For analysis of European firms' involvement in the Congo War at the height of the coltan price boom, see Jeroen Cuvelier and Tim Raeymaekers' *Supporting the War Economy in the DRC: European companies and the coltan trade* and *European companies and the coltan trade: an update*, both published by the International Peace Information Service (IPIS) (Ghent, Belgium, 2002), and Custers, Raf, Cuvelier, Jeroen, and Verbruggen, Didier, Culprits or Scapegoats? Revisiting the role of Belgian mineral traders in eastern DRC (Antwerp, IPIS, 2009).

Chapter 4 relied on a mix of primary and secondary material about coltan initiatives. The United Nations reports and resolutions referred to in chapter 5 are all available at www. un.org. For a discussion of bias by the United States, French and British governments and how this might have shaped UN reports, see François Grignon, Economic Agendas in the Congolese Peace Process. In: Michael Nest, with François Grignon and Emizet F. Kisangani, *The Democratic Republic of Congo: Economic dimensions of war and peace* (Boulder CO, Lynne Rienner, 2006), pp. 84–9. For analysis of the technical challenges of the proposed fingerprinting scheme for tantalum, see Frank Melcher, Torsten Graupner, Friedhelm Henjes-Kunst, Thomas Oberthür, Maria A. Sitnikova, Eike Gäbler, et al., *Analytical Fingerprint of Columbite-Tantalite (Coltan) Mineralisation in Pegmatites – Focus on Africa*. Paper for the Ninth International Congress for Applied Mineralogy, Brisbane, 8–10 September 2008; Frank Melcher, Maria A. Sitnikova, Torsten Graupner, Nicola Martin, Thomas Oberthür, Friedhelm Henjes-Kunst, et al., Fingerprinting of conflict minerals: columbite-tantalite ('coltan') ores. *SGA News* 23, 2008, Society for Geology Applied to Mineral Deposits, p. 1. For information on the Enough Project and its activities, see www.enoughproject.org and www.raisehopeforcongo.org. These websites have links to video testimonies by celebrities and other campaign videos that have been produced in support of the Enough Project's conflict minerals campaign, as well as useful information about US conflict minerals legislation. For information about Friends of the Congo and its Congo Week and Breaking the Silence activities, see www. friendsofthecongo.org.

Global initiatives focused on reforming production and trade of natural resources all have their own websites. For information about the Kimberley Process Certification Scheme (KPCS), see www.kimberleyprocess.com. For a summary of criticisms of the KPCS, see Nicolas Garrett and Harrison Mitchell, *Trading Conflict for Development: Utilising*

the trade in minerals from eastern DR Congo for development
(Aston Sandford, UK, Resource Consulting Service, and
Washington DC, Communities and Small-Scale Mining initiative, 2009). For information on the OECD guidelines, see
www.oecd.org and the OECD Guidelines for Multinational
Enterprises: Text, commentary and clarifications (Paris,
Organisation for Economic Co-Operation and Development,
2001). For information on the Global Compact, see www.
unglobalcompact.org. See http://eiti.org for information
about the Extractive Industries Transparency Initiative. For
a critique of the EITI, see Ivar Kolstad and Arne Wiig's Is
Transparency the Key to Reducing Corruption in Resource-
Rich Countries? World Development, 37(3), 2009, 521–32. For
information about the Publish What You Pay campaign, see
www.publishwhatyoupay.org; for the International Council
on Minerals and Mining's sustainable development framework see www.icmm.com; for the Initiative for Responsible
Mining Assurance see http://responsiblemining.net.

For analysis of the tactics of, and challenges facing, trans-
national activism see Margaret E. Keck and Kathryn Sikkink,
Activists Beyond Borders (Ithaca NY, Cornell University
Press, 1998); Clifford Bob, The Marketing of Rebellion:
Insurgents, media and international activism (New York,
Cambridge University Press, 2005); Sidney Tarrow, The New
Transnational Activism (New York, Cambridge University
Press, 2006); Andrew Fenton Cooper, Celebrity Diplomacy
(Boulder, CO, Paradigm, 2008). For excellent analyses of
transnational activist campaigns that had an impact on the
Congo, see Adam Hochschild's King Leopold's Ghost: A story of
greed, terror, and heroism in colonial Africa (Boston, Houghton
Mifflin, 1998) and Bury the Chains: Prophets and rebels in
the fight to free an empire's slaves (New York, Mariner Books,
2005). For a discussion of the constraints on transnational
activism in developing countries, including China, see Nicole

Piper and Anders Uhlin (eds.), *Transnational Activism in Asia: Problems of power and democracy* (London, Routledge, 2004); Peter Ho and Richard Edmonds (eds.), *China's Embedded Activism: Opportunities and constraints of a social movement* (London, Routledge, 2007); Gay W. Seidman, *Beyond the Boycott: Labor rights, human rights, and transnational activism* (New York, Russell Sage Foundation, 2007).

For analysis of relations between the DRC and China, and Africa and China, see Barry Sautman and Hairong Yan, Friends and Interests: China's Distinctive Links with Africa, *African Studies Review* 50(3), 2007, 75–114; Chris Alden, *China in Africa* (London, Zed Books, 2007); Hannah Edinger and Johanna Jansson, China's 'Angola Model' comes to the DRC, *China Monitor* 34, 2008, 4–6; Daniel Large, Beyond 'Dragon in the Bush': The study of China–Africa relations, *African Affairs* 107(426), 2008, 45–61. The Trade Law Centre for Africa (Stellenbosch, South Africa) publishes China–Africa trade data on its website (www.tralac.org).

Index